Conservin
Agricultur

Conserving Agriculture

Editor
Dr. Ajoy Gangopadhyay

2025

GENE-TECH BOOKS

ISBN 978-81-89729-68-4

Published by: **GENE-TECH BOOKS®**
4762-63/23, Ansari Road, Darya Ganj,
New Delhi - 110 002
Phone: 09899295259, 011-43003222
E-mail: genetechbooks@yahoo.co.in

PRINTED IN INDIA

Preface

Conservation agriculture refers to a range of soil management practices that minimise effects on composition, structure and natural biodiversity and reduce erosion and degradation. Such practices include direct sowing/ no-tillage, reduced tillage/ minimum tillage, non or surface- incorporation of crop residues and establishment of cover crops in both annual and perennial crops. As such the soil is protected from rainfall erosion and water runoff; the soil aggregates, organic matter and fertility level naturally increase, and soil compaction is reduced. Furthermore, less contamination of surface water occurs, water retention and storage is enhanced, which allows recharging of aquifers.

Conservation agriculture holds tremendous potential for all sizes of farms and agro-ecological systems, but its adoption is perhaps most urgently required by smallholder farmers, especially those facing acute labour shortages. It is a way to combine profitable agricultural production with environmental concerns and sustainability and it has been proven to work in a variety of agroecological zones and farming systems.

Conservation agriculture techniques are being progressively adopted under different climatic conditions, from cold regions to tropical regions, semi-arid areas to high rainfall areas. Farmers, advisers and agricultural administrators need to be aware and further informed of the agronomic, economic and environmental benefits of these new developments and progress and therefore an aim is to encourage interaction between farmers, advisors and scientists throughout the developed and developing world.

The purpose of this book is to show how conservation agriculture can increase crop production while reducing erosion and reversing soil fertility decline, improving rural livelihoods and restoring the environment. It will serve as invaluable source of references to the scientists, students and teachers working in public and private sectors in the discipline of agriculture.

Editor

Contents

1

Conservation Agriculture

Conservation Agriculture (CA) is a concept for resource-saving agricultural crop production that strives to achieve acceptable profits together with high and sustained production levels while concurrently conserving the environment. CA is based on enhancing natural biological processes above and below the ground. Interventions such as mechanical soil tillage are reduced to an absolute minimum, and the use of external inputs such as agrochemicals and nutrients of mineral or organic origin are applied at an optimum level and in a way and quantity that does not interfere with, or disrupt, the biological processes. CA is characterised by three principles which are linked to each other, namely:

— Continuous minimum mechanical soil disturbance.

— Permanent organic soil cover.

— Diversified crop rotations in the case of annual crops or plant associations in case of perennial crops.

Conventional "arable" agriculture is normally based on soil tillage as the main operation. The most widely known tool for this operation is the plough, which has become a symbol of agriculture. Soil tillage has in the past been associated with increased fertility, which originated from the mineralisation of soil nutrients as a consequence of soil tillage. This process leads in the long term to a reduction of soil organic matter. Soil organic matter not only provides nutrients for the crop, but it is also, above all else, a crucial element for the stabilisation of soil structure.

Therefore, most soils degrade under prolonged intensive arable agriculture. This structural degradation of the soils results in the formation of crusts and compaction and leads in the end to soil erosion. The process is dramatic under tropical climatic situations but can be noticed all over the world. Mechanisation of soil tillage, allowing higher working depths and speeds and the use of certain implements like ploughs, disk harrows and rotary cultivators have particularly detrimental effects on soil structure.

Soil erosion resulting from soil tillage has forced us to look for alternatives and to reverse the process of soil degradation. The logical approach to this has been to reduce tillage. This led finally to movements promoting conservation tillage, and especially zero-tillage, particularly in southern Brazil, North America, New Zealand and Australia. Over the last two decades the technologies have been improved and adapted for nearly all farm sizes; soils; crop types; and climatic zones. Experience is still being gained with this new approach to agriculture and FAO has supported the process for many years.

Experience has shown that these techniques, summarised as conservation agriculture (CA) methods, are much more than just reducing the mechanical tillage. In a soil that is not tilled for many years, the crop residues remain on the soil surface and produce a layer of mulch. This layer protects the soil from the physical impact of rain and wind but it also stabilises the soil moisture and temperature in the surface layers. Thus this zone becomes a habitat for a number of organisms, from larger insects down to soil borne fungi and bacteria.

These organisms macerate the mulch, incorporate and mix it with the soil and decompose it so that it becomes humus and contributes to the physical stabilisation of the soil structure. At the same time this soil organic matter provides a buffer function for water and nutrients. Larger components of the soil fauna, such as earthworms, provide a soil structuring effect producing very stable soil aggregates as well as uninterrupted macropores leading from the soil surface straight to the subsoil and allowing fast water infiltration in case of heavy rainfall events.

This process carried out by the edaphon, the living component of a soil, can be called "biological tillage". However, biological tillage is not compatible with mechanical tillage and with increased mechanical tillage the biological soil structuring processes will disappear. Certain operations such as mouldboard or disc ploughing have a stronger impact on soil life than others as for example chisel ploughs.

Most tillage operations are, however, targeted at loosening the soil which inevitably increases its oxygen content leading in turn to the mineralisation of the soil organic matter. This inevitably leads to a reduction of soil organic matter which is the substrate for soil life. Thus agriculture with reduced, or zero, mechanical tillage is only possible when soil organisms are taking over the task of tilling the soil. This, however, leads to other implications regarding the use of chemical farm inputs. Synthetic pesticides and mineral fertiliser have to be used in a way that does not harm soil life.

As the main objective of agriculture is the production of crops, changes in the pest and weed management become necessary with CA. Burning plant residues and ploughing the soil is mainly considered necessary for phytosanitary reasons: to control pests, diseases and weeds. In a system with reduced mechanical tillage based on mulch cover and biological tillage, alternatives have to be developed to control pests and weeds. Integrated Pest Management becomes mandatory.

One important element to achieve this is crop rotation, interrupting the infection chain between subsequent crops and making full use of the physical and chemical interactions between different plant species. Synthetic chemical pesticides, particularly herbicides are, in the first years, inevitable but have to be used with great care to reduce the negative impacts on soil life. To the extent that a new balance between the organisms of the farm-ecosystem, pests and beneficial organisms, crops and weeds, becomes established and the farmer learns to manage the cropping system, the use of synthetic pesticides and mineral fertiliser tends to decline to a level below that of the original "conventional" farming system.

Conservation Agriculture, understood in this way, provides a number of advantages on global, regional, local and farm level:

— It provides a truly sustainable production system, not only conserving but also enhancing the natural resources and increasing the variety of soil biota, fauna and flora (including wild life) in agricultural production systems without sacrificing yields on high production levels. As CA depends on biological processes to work, it enhances the biodiversity in an agricultural production system on a micro- as well as macro level.

— No till fields act as a sink for CO_2 and conservation farming applied on a global scale could provide a major contribution to control air pollution in general and global warming in particular. Farmers applying this practice could eventually be rewarded with carbon credits.

— Soil tillage is among all farming operations the single most energy consuming and thus, in mechanised agriculture, air-polluting, operation. By not tilling the soil, farmers can save between 30 and 40% of time, labour and, in mechanised agriculture, fossil fuels as compared to conventional cropping.

— Soils under CA have very high water infiltration capacities reducing surface runoff and thus soil erosion significantly. This improves the quality of surface water reducing pollution from soil erosion, and enhances groundwater resources. In many areas it has been observed after some years of conservation farming that natural springs that had dried up many years ago, started to flow again. The potential effect of a massive adoption of conservation farming on global water balances is not yet fully recognised.

— Conservation agriculture is by no means a low output agriculture and allows yields comparable with modern intensive agriculture but in a sustainable way. Yields tend to increase over the years with yield variations decreasing.

— For the farmer, conservation farming is mostly attractive because it allows a reduction of the production costs, reduction of time and labour, particularly at times of peak demand such as land preparation and planting and in mechanised systems it reduces the costs of investment and maintenance of machinery in the long term.

Disadvantages in the short term might be the high initial costs of specialised planting equipment and the completely new dynamics of a conservation farming system, requiring high management skills and a learning process by the farmer. Long term experience with conservation farming all over the world has shown that conservation farming does not present more or less but different problems to a farmer, all of them capable of being resolved.

Under forest, the great production and cycling of foliage results in much biological activity, humus formation, and hence a dark coloured topsoil. Because of the great numbers of insects and worms there are large pores, which allow water infiltration. In contrast, under annual crops, leaf production is much less, the biomass is largely removed, the soil is tilled several times each year and so is much drier. Consequently, less food and moisture are available for earthworms and other insects, and their habitat is repeatedly disturbed or destroyed.

Where topsoil has been eroded, and soil layers of poorer quality for root growth have become exposed, it is essential to rehabilitate and restore the soil to bring it up to good productive capacity for the next crop or pasture. Failing this, a spiral of degradation is set in motion as a result of the reduced vegetative cover and biomass production and reduced soil and water retention. Thus the quality of the soil that is left behind should be of even greater concern than the quantity and quality of that which has been lost.

Farmers need to create favourable conditions for soil life and should manage organic matter so as to create a fertile soil in which healthy plants can develop. In tropical rainfed agriculture, in which poor farmers generally suffer from decreasing soil fertility and declining soil water dynamics, the restoration of soil organic matter is essential for the stabilisation of production. However, this cannot be accomplished by merely incorporating organic matter into the soil, as under tropical conditions, the degradation process is too fast to allow any medium or long-term improvement of soil properties. Moreover, incorporation implies tilling the soil, which accelerates organic matter breakdown and destroys soil structure and organisms.

The primary need is to feed soil organisms (bacteria, fungi, earthworms, etc.) and to regulate their living conditions, while protecting them from chemical and mechanical impacts. For example, shallow tillage, ridge-tillage, or zero-tillage and surface management of crop residues has often led to increases in earthworm activity compared to areas where deep tillage is practised. Providing a permanent or semi-permanent soil cover (growing crops, crop residues or mulch) provides food for soil organisms, protects the soil from the destructive forces of rain, wind and sun, improves water infiltration, reduces soil moisture loss, and regulates the soil microclimate.

This practice should be accompanied by others related to conservation agriculture, which intend to minimise soil disturbance and protect and nourish the soil life, such as:

— reducing or eliminating tillage operations;

— practising crop rotations;

— using fertilisers as appropriate;

— relying on integrated pest and weed management.

The benefits of conservation agriculture include agro-environmental features. Nutrient losses may be minimised through the appropriate use of deep-rooting cover crops that recycle nutrients leached from the topsoil, moisture management, and improved collection, storage and application of wastes from crops, livestock and the household (food wastes). Nutrients that are harvested and removed may be replaced through symbiotic nitrogen fixation, organic matter from elsewhere, or the complementary use of fertilisers and feed supplements.

Agro-environmental features of conservation agriculture

- — Soil loss does not exceed rates of soil formation;
- — Soil fertility and soil structure are maintained or enhanced;
- — Biodiversity is maintained or enhanced;
- — Downstream effects of run-off or leaching do not impair water quality;
- — Rainfall is managed to avoid excessive runoff;
- — Emissions of greenhouse gases are reduced;
- — Food production levels are maintained or enhanced;
- — Environmental stewardship is engendered amongst rural communities and producers of all types, ensuring continuity of sound land management.

Pest management can also benefit from conservation practices that enhance biological activity and diversity, and hence competitors and predators, as well as alternative sources of food. For instance, most nematode species (especially the pathogens) can be significantly reduced by application of organic matter, which stimulates the action of several species of fungi attacking nematodes and their eggs.

The current concept shift from soil being a thin layer of material at the outside of the lithosphere immediately below the atmosphere to a living entity that has dynamics of root growth and soil fauna, temperature, moisture and oxidation-reduction, has profound significance for ecological study and practical management. Nutrients that are lost from the soil by crop production, erosion and leaching need to be replaced and the availability of all nutrients needs to be optimised. The broader focus of conservation agriculture embraces not only the nutrient content of soils but also their structure and biological status, which are determinants of sustained productivity.

In many cases this may require a combination of changes in tillage and soil management practices, crop rotations and planting times, soil conservation measures, the strategic use of organic materials and the appropriate use of inorganic fertilisers to match farmers' combinations of crops, land, availability of organic materials and market opportunities. An improved approach to the integrated and sustainable use of natural resources requires a paradigm centred on the user's role and the significance of the soil's biological and architectural dynamics, both at and below the surface, as much as on the increased synergy between local, internal and external forces. As farmers use management skills and better knowledge to work more closely with the biological world, they will often find ways to reduce purchases of external inputs.

With a new emphasis on conservation agriculture has come a reawakening of interest in soil organic matter. Some issues, such as soil fertility (thus food security), water storage, compaction, and erosion are directly related to soil organic matter. Others, such as disease and insect pest infestations, may be indirectly related to it. Thus, the build-up and maintenance of the soil biota and good levels of organic matter in soils are of critical importance.

The adoption of conservation agriculture requires the opening up of dense and compacted soils as well as an opening of minds and innovative thinking. For example, almost all of the past limitations to change in Brazil have been overcome with positive and creative thinking. During 1998 and 1999, 140 extensionists were trained from seven different states. The training courses were a great success and completely turned around the attitudes of extension services to zero tillage, paving the way for collaboration in more pilot projects with small farmers and leading to considerable benefits to the small farm sector.

Agricultural science generally has poorly understood, overlooked or ignored indigenous knowledge and traditional approaches. Soil conservation staff commonly have focused on what they have seen as technically desirable solutions to problems of erosion and runoff. Extension agencies often find it difficult to learn from farmers and rural people. Few systemic processes exist for enhanced two-way feedback on performance. An examination of the situation in tropical Brazil identified several reasons for the delay in research attention to farmers' practices.

Land Degradation and Improvement of Land Productivity

The control of soil erosion and establishing permissible amounts of soil loss have long been principal foci in addressing land degradation and aiming at increasing and stabilising agricultural production. In contrast to this narrow approach to soil-related problems, it is being increasingly recognised that land and its soil components should be looked upon as a living resource to be nurtured and used in sustainable and responsible ways.

In the past, soil conservation has been advocated as a necessary starting point to raise crop yields. Soil erosion has conventionally been perceived as one of the main causes of land degradation and the main reason for declining yields in tropical regions. Based on these assumptions, conservation measures were directed at three main components:

— physical works to catch, guide and prevent damage by run-off;

— pressures to stop people from deforesting the area and to reduce the number of grazing animals;

— planning of different land uses according to Land Use Capability Classifications, based on the assessment of different degrees of erosion hazards.

Experience has shown that none of the recommended physical and institutional anti-erosion methods was widely adopted by the smallholder farmers of tropical regions. Since conserving soil does not by itself raise yields, and is not the farmers' overriding concern (while improving productivity may be), it is advisable to emphasise those practices of good soil and crop management that have, positive effects on conservation. This insight has led to a switch from stopping erosion to assisting farmers to achieve a more conservation-effective, higher and more stable production.

For examle in Southern Brazil, large areas of arable land suffered from erosion to such an extent that the very livelihood of the farmers was being endangered. Initial efforts to contain the damage by the implementation of conservation works such as terracing did not prove effective. As research studies developed, scientists confirmed that the erosion problem was due to the way the land between terrace banks was managed. Even if the terraces were well constructed, the rate of rainwater infiltration was progressively reduced due to excessive soil movement and compaction. The technique presented as a solution when used as an isolated practice, i.e. the construction of terraces, accentuated rather than alleviated the problem.

This resulted in the revival of the ancient practice of green manuring. Firstly with the clear objective of erosion control, which later developed into what could be defined as good soil management. More important than using physical barriers to control runoff, which is responsible for only 5 percent of erosion, research showed that the ideal solution is to maintain soils covered as much of the time as possible with growing plants or crop residues. By avoiding the detachment of soil particles by raindrop impact, which accounts for 95 percent of erosion, soil losses are avoided and at the same time the soil can be cultivated in conditions similar to those found in forests.

This was accompanied by the emergence of new systems of land preparation such as minimum tillage and direct sowing techniques as alternatives to the conventional practices introduced from temperate climates. Depending on the crop to be sown, the area of soil to be disturbed is limited to a narrow strip, between 10 and 50 cm wide. In this strip, the vegetative cover is partially incorporated and the soil surface is still 60-80 percent protected from raindrop impact and the sun's rays. Direct sowing consists of the elimination of ploughing or soil disturbance using traditional equipment such as the plough or cultivator. Direct sowing is practised through a cover of crop residues or in a narrow partially cleared strip.

Similarly, especially in the arid and semi-arid tropics, it is opportune to emphasise with farmers the management of rainwater as a productive resource rather than merely as a means of saving soil. Achieving better infiltration and in-soil storage of rainwater when these have been limitations, while favouring agricultural production, automatically also reduces soil and water movement and transport. In this regard, to enhance water availability and retain soil productivity it is important to consider those practices which promote rainfall capture in the soil before considering those which aim to control run-off - they are complementary in a sequence, and are not competing alternatives.

In areas of high rainfall and tendency to soil water-logging, conservation of water and soil requires careful management of soil structure and the vegetative cover to enhance infiltration and maintain above-ground and internal drainage. Facilitating farmers to improve their land care - land husbandry - thus provides a more effective response than efforts to combat erosion alone. It specifically recognises farmers' desire to raise yields and incomes as they stabilise or reverse resource depletion. It also provides opportunities for governments to harmonise certain national objectives (better management of natural resources and development of sustainable agriculture) with major objectives of farm families (secure livelihoods). However, this approach requires many adjustments in common thinking.

In Lempira (Honduras), farmers moved from a traditional slash and burn system to the Quesungual system: conservation agriculture with an agroforestry component. An economic analysis of this transition showed that during the first two years maize and sorghum yields are about equal to those obtained with the traditional slash and burn system. From the third year, however, their yields increase, in addition to which the plot provides the farmer with firewood and posts, which give an extra value to the production.

Because of the increased production of maize, the quantity of stover increased as well; this can be sold as livestock fodder. Additionally, from the first year onwards the farmer can rent out the land for livestock grazing, because of the increased biomass production. Usually this is done for two months. The application of the Quesungual system not only meets the household subsistence needs for fruit, timber, firewood and grains, but generates a surplus, which generates an extra income when sold in the market.

Other benefits quickly appreciated by farmers include the reduction of the amount and costs of labour and energy required for land preparation and sowing, due to the fact that the soil becomes soft and easy to work. Ploughing the soil is by far the most energy- and time-consuming operation for the farmer. In many farming systems, it constitutes an important bottleneck, often due to the necessity to hire equipment which does not arrive in time and therefore delays planting.

Of the total energy used in crop production in North Africa in 1987, 69 percent was derived from people, 17 percent from animals, and 14 percent from tractors. In sub-Saharan Africa this ratio was 89:10:1. Findlay and Hutchinson estimated that 80-100 person-days/ha would be needed to prepare a land for planting with hand hoes. Animal-drawn mouldboard ploughing may take two or three days, whereas tractor ploughing may require only two or three hours.

Although it is often recommended that farmers should plough immediately after harvest, most farmers wait until the first rains before commencing seedbed preparation. Because the majority of African farmers have no direct access to animal or motorised traction, seedbeds are often prepared too late, the cropping season shortened, and crop yields reduced.

Under conservation agriculture, in most systems only a small proportion of the land is worked instead of ploughing or hoeing the whole area to be planted. Cultivation is also usually shallower than conventional tillage. Herbicides may be used in some systems, hand hoes in others, and farmers who have animal-drawn ploughs can fit simple and inexpensive tines or subsoilers. Farmers using conservation tillage reduced the production costs of soybeans per hectare by US$67 in Argentina, by US$35 in the USA and by US$27 in Brazil. The farmers' point of view is a central consideration in an adoption process; they will not change their practices if they do not see any benefit. In fact, the reductions in costs and time required are usually the most compelling reasons for farmers to adopt conservation tillage.

Among the benefits farmers found in applying conservation agriculture practices within the Quesungual system, were:

— improved soil moisture conservation, which permits a good development of the crop, even in very bad conditions;

— less soil erosion;

— reduced disease incidence in the bean crop due to the mulch;

— production of firewood and fruits from the trees and shrubs; timber trees can also be cut after about 7 years and used for construction or sold;

— the soil becomes more fertile and the effect of fertilisers on the production is higher;

— agricultural production is higher than in traditionally managed plots;

— plots can be cultivated for longer periods than in the slash and burn system;

— less labour involved in the establishment and maintenance of the system;

— improved soil workability, which implies less labour need during land preparation and sowing;

— harvested products, such as beans and maize, can be dried by hanging them over the tree trunks;

— cattle can feed on the residues after the harvests of maize and sorghum;

— trees and shrubs in the plot provide shade for the farmer and attract animals and insects, birds and butterflies.

The disadvantages mentioned by the same farmers were:

— equal or slightly lower grain production during the first year compared to the traditional system;

— higher incidence of slugs in the bean crop during the first years;

— sometimes too much soil cover impedes the germination of the seeds;

— the shade can result in higher disease incidence during periods of high rainfall because of higher humidity.

Environment Benefits

Experience has shown that conservation agriculture systems achieve yield levels as high as comparable conventional agricultural systems but with less fluctuations due, for example, to natural disasters such as drought, storms, floods and landslides. Conservation agriculture therefore contributes to food security and reduces risks for the communities (health, conditions of living, water supply), and also reduces costs for the State (less road and waterway maintenance, less emergency assistance).

Conservation agriculture also contributes to wider environmental benefits such as:

— improved management of soil and water resources from farm to watershed levels: less flooding, less erosion, less desertification, more constant flow in the rivers, better recharge of groundwater resources, improved water quality (less pollution) and reduced siltation effects downstream;

— increased carbon sequestration and less carbon release (less fuel used, less organic matter degradation);

— increased biodiversity through diversification.

Management Practices on Soil Fauna and Soil Fertility

The greater production of biomass in a system with cover crops and zero or reduced tillage compared to monocrop cultures with conventional tillage, leaves a protective blanket of leaves, stems and stalks from the previous crops on the surface. In this way organic matter can be built up in the soil, which has great influence on the activity and the population of the micro-organisms. This results in a greater biological activity, more humus formation, and hence a darker coloured topsoil. With time, in reduced or zero tillage systems, soil fauna take over the functions of traditional soil tillage, which is loosening the soil and mixing the soil components. In addition, the increased biological activity creates a stable soil structure through accumulation of organic matter.

A component of soil quality maintenance is favouring the activity of beneficial soil organisms. Among the most important species are the root nodule bacteria involved in biological nitrogen fixation. Several studies have indicated that zero tillage systems increase soil microbial biomass and the size of the microbial population. For zero tillage systems in southern Brazil, differences of about 50 percent in soil biomass and rhizobial populations compared to conventional tillage were reported. Evaluations have demonstrated that some crop rotations and zero tillage favour bradyrhizobia populations, nodulation and thus nitrogen fixation and yield.

The vegetal cover on the soil surface creates a more humid environment, which is conducive to the activity of soil organisms. The greater numbers of worms, termites, ants and millipedes combined with a higher density of plant roots result in more large pores, which in turn increase water infiltration. Thus in an experiment in southern Brazil rainwater infiltration increased from 20 mm/h under conventional tillage to 45 mm/h under no-tillage. As a result soil erosion may be reduced to a level below the regeneration rate of the soil and the groundwater resources may be maintained or even enhanced. Leaching of plant nutrients or other substances into the aquifer is also reduced compared to conventional arable agriculture. Soil moisture storage and availability is also improved both by soil cover (less evaporation, more infiltration) and soil organic matter. All these phenomena are improving plant nutrition.

Reduction of Water Pollution

The change in land use and management associated with conservation-effective practices leads to a significant reduction in erosion, and thus to a reduction in water pollution and contamination. Indicators that can be used to measure this reduction of water pollution include:

— Water turbidity and the concentration of sediments in suspension;

— Total sediment loss and associated loss of nutrients;

— Reduction of water treatment costs.

The reductions varied between 50 and 80 percent, depending on soil types predominating in the areas. These reductions are due to increased perennial crops (banana and pasture) on hillsides, thus reducing the erosion process. Total sediment loss decreased by 16 percent and the cost of plant nutrients by 21 percent. Reduced sediment loss and less soil particles in suspension also reduced the cost of water treatment.

Development of Biodiversity

The result of increasing soil cover, through living crops and crop residues, is an increase in the variety and variability of animals, plants and micro-organisms, which are necessary to sustain key functions of the agro-ecosystem. Conservation agriculture provides more habitats for birds, small mammals, reptiles and earthworms, amongst others, and more food, including insects and seeds, which in turn leads to an increase in species and population. The increased production from conservation-based agriculture also makes it possible to set aside areas for natural regeneration.

Traditionally, livestock was produced in hilly areas, in an extensive form, without any management of the resources, leading to erosion and environmental problems. The solution for livestock production was to change towards a more intensified production system, with the objective to reduce degradation risk, improve the nutrition status of the cattle and liberate areas which can be used for other activities, including the natural regeneration of the vegetation.

The intensified system is based on semi-stall-fed production of livestock and was initiated through a farm planning in order to define land use capacity and to select the areas most suitable for livestock raising. Part of these areas was then sown with improved pastures and the rest with fodder crops. The improved pastures were divided into smaller parcels to permit rotational grazing. The fodder crop area is fertilised with manure produced in small stables.

The intensification of the livestock system has resulted in spectacular increases in the production of meat and milk. The relocation of livestock activities has led to the natural regeneration of severely eroded land and areas unsuitable for agricultural production, which is having a positive effect on the biodiversity and allows the government to "trade" the area in international treaties dealing with forest protection or carbon sequestration.

2

Adoption of Conservation Agriculture

Conservation agriculture is more profitable in steep-sloping, high rainfall tropical regions than in flatter temperate areas, since the former would be subject to a higher risk of erosion under conventional tillage. But such a generalisation would hide a number of the complexities that make the analysis of financial returns from CA difficult.

Temperate Agro-ecological Zone

One of the first comprehensive financial analyses of CA on large farms in developed countries compared the on-farm costs of conventional tillage with conservation tillage in the United States. CA has a small cost advantage over conventional tillage but that site-specific conditions could alter this result in various ways (Table 1). The following input cost aspects form the basis for these general conclusions.

Machinery and Fuel Costs

This is the most important cost item for larger producers and so the impact of CA on these expenditure items is critical. Most analyses suggest that CA reduces machinery costs. Zero or minimum tillage means that farmers can use a smaller tractor and make fewer passes over the field. This also results in lower fuel and repair costs. However, this simple view masks some complexities in making a fair comparison. For example, farmers may see CA as a complement to rather than as a full substitute for their existing practices. If they only partially switch to CA, then their machinery costs may rise as they must now provide for two cultivation systems, or they may simply use their existing machinery inefficiently on their CA fields.

Table 1. Comparison of conventional and conservation tillage costs for maize and soybeans in the United States, 1979 and 1992

Crop/ cost item	Per acre costs in 1979			Per acre costs in 1992		
	(1) Conve-ntional tillage	(2) Conser-vation tillage	Ratio (1/2)	(3) Conven-tional tillage	(4) Conser-vation tillage	Ratio (3/4)
Maize						
machinery & fuel	45.34	38.34	1.18	55	37-44	1.36
pesticides	1/8.72	11.63	0.75	10-15	5-25	0.83
labour	13.24	6.62	2.00	8	5-7	1.33
total selected costs	67.30	56.59	1.19	73-78	56-76	1.14
Soybeans						
machinery & fuel	38.11	33.11	1.15	55	37-44	1.36
pesticides	1/9.13	12.17	0.75	14-28	7-40	0.89
labour	12.21	6.10	2.00	8	5-7	1.33
total selected costs	59.45	51. 38	1.16	77-91	58-91	1.13

To capture such complexity, economists distinguish between short-run and long-run costs, where the former assumes no adjustment to existing capital equipment and the latter assumes such an adjustment. Similarly, the expectation is for fuel costs to be lower under CA, and this is generally the finding in most studies. Falling fuel prices should encourage greater adoption of CA. One study shows that the price of crude oil has a statistically significant but relatively minor effect on the intensity of CA. It finds that a 10 percent increase in the United States in the price of crude oil is associated with an expansion in planted hectares under CA of 0.4 percent, with the expansion being concentrated primarily on existing CA farms.

Pesticide Costs

Offsetting lower machinery costs are higher herbicide applications under CA, especially during the early adoption period and with no-till. Indeed, herbicides substitute for the use of machinery to keep weeds under control. Site-specific factors are important as perennial weeds can present problems for CA. Nonetheless, herbicide application rates and the ability to fully control weeds under CA in all situations remains a controversial and continuing area of CA research. Recent assessments have tended to argue that herbicide applications decline over time and may eventually fall to a level equal to that of conventional tillage. Insect control is less an issue in conventional and CA comparisons. As most pesticides are petroleum based, crude oil prices are liable to affect their cost to farmers. If so, then a higher crude price would mean higher herbicide costs, partially offsetting CA's relative cost advantage stemming from lower machinery fuel requirements.

Labour Costs

Much attention has focused on the apparent reduction in labour requirements under CA. This reduction follows from the decreased demand for labour for land preparation at the beginning of the growing season. Some estimates put this reduction at 50-60 percent during this time period. On large mechanised farms in the developed world the true impact of this saving is small as labour costs account for under 10 percent of total per acre costs (Table 1). However, on some farms in the developed world, the trend towards increased off-farm work has made even the relatively small labour savings under CA attractive. Indeed, some case studies have cited the time savings provided by CA as the primary motivation for the adoption of conservation tillage.

Fertiliser and other Input Costs

Most comparative analyses of the costs of conventional tillage versus conservation tillage assume that other production inputs remain unchanged following a switch to CA. A debate continues concerning fertiliser use under CA as there is evidence, that CA adoption affects nitrogen use by crops and leaching. Uri finds some increase in fertiliser use by maize farmers adopting conservation tillage in the United States. Additionally, if the application of fertilisers under CA requires greater management skill, then application costs could rise even if application rates do not. A more general finding is that CA requires greater management skills and it may be costly for farmers to acquire these. CA may also affect seed purchases as farmers may be able to avoid some pest problems by investing in more resistant seed varieties. However, this increases costs.

The comparative data in Table 1 reveal a consistent picture in recent decades concerning conservation tillage costs in the United States. More recent estimates tend to show a wide range for CA, recognising the variation in site-specific conditions. Perhaps more significantly, the cost items listed in Table 1 represent only a subset of total costs as other production inputs and land were assumed to remain constant under either cultivation system. Putting the cost savings attributable to CA in the context of these total costs, any cost advantage amounts to about 5-10 percent in 1979 and probably about the same in the 1990s.

Also missing from many cost comparisons of conventional and conservation tillage is an analysis of risk factors. One aspect of risk is a recognition that yields might vary under the different cultivation systems. As the results for temperate climates are often contradictory, and any differences are usually not statistically significant, most analysts simply assume no change in yield. Similarly, the impact of adopting CA on yield variability and risk is controversial. Some studies argue that CA increases yield variability in many situations, thereby worsening risk. By contrast, Australian research shows a reduced variability in crop yields with CA, while work in Canada indicates that the net returns were higher under CA than conventional practices in bad years, but lower when averaged over time. Firm conclusions on whether risk is increased or reduced under CA remain elusive.

More certain are the impacts of CA on cropping intensity. With reduced field preparation time, the cropping cycle is shorter, allowing more crops in a given period and even double cropping where it was not possible previously. Where this benefit is available from CA, more efficient utilisation of the fixed land resource results in higher annual net returns per hectare. Moreover, farmers may adjust their cropping strategy when switching to CA. Hence, yield trials comparing the same crop under either cultivation system may not represent reality. In fact, fully adopting CA involves switching to a suitable crop rotation that will probably differ from the conventional cropping strategy used previously. For this reason, some writers have called for a broader whole farm approach to comparative assessments in temperate agriculture.

Overall, a comparison between conventional and conservation practices in temperate agro-ecological zones hinges on two offsetting effects. One involves CA's labour and possibly machinery cost savings, while the other involves higher herbicide costs, at least initially, under CA. Depending upon the magnitude of each of these effects, CA may appear either more or less costly. For example, in Saskatchewan, Canada, researchers found that the higher herbicide costs characterising CA overwhelmed any cost savings associated with labour, fuel, machine repair and overheads. Similarly, Stonehouse and Bohl used a linear programming model to argue that conservation tillage in a cash-crop farm system is not profitable. However, most developed-country studies reviewed find that CA demonstrates at least minor cost savings over conventional practices. However, these savings have not been sufficient to induce adoption by large numbers of farmers on large mechanised farms. These farmers may resist new practices unless there is a promise of much higher financial returns.

Tropical/Temperate Agro-ecological Zone

Large-scale mechanised farming is common in many parts of Latin America and farmers have adopted CA on large portions of this cultivated area. While most of the comparative cost analysis presented above for temperate northern regions would apply here, the advantage of CA in Latin America has been more pronounced. In part, this greater advantage reflects physical and climate factors, but also the differences in the nature of the technology adopted.

In Paraguay, yields under conventional tillage declined 5-15 percent over a period of ten years, while yields from zero tillage increased 5-20 percent. Savings in fertiliser and herbicide inputs dropped by an average of 30-50 percent over the same period. In Brazil, over a 17-year period, maize and soybean yields increased by 86 and 56 percent, respectively, while fertiliser inputs for these crops fell by 30 and 50 percent, respectively. In addition, soil erosion in Brazil fell from 3.4-8.0 t/ha under conventional tillage to 0.4 t/ha under no-till, and water loss fell from approximately 990 to 170 t/ha.

As a result, the financial benefits for farmers in Latin America who have adopted CA have been striking. However, these take time to fully materialise. Sorrenson compared the financial profitability of CA on 18 medium and large-sized farms with conventional practice in two regions of Paraguay over 10 years. He found that by the tenth year net farm income had risen on the CA farms from under US$10 000 to over US$30 000, while on conventional farms net farm income fell and even turned negative. Medium and large-scale farmers have experienced:

- less soil erosion, improvements in soil structure and an increase in organic matter content, crop yields and cropping intensities;
- reduced .time between harvesting and sowing crops, allowing more crops to be grown over a 12-month period;
- decreased tractor hours, farm labour, machinery costs, fertiliser, insecticide, fungicide and herbicide, and cost savings from reduced contour terracing and replanting of crops following heavy rains;
- lower risks on a whole-farm basis because of higher and more stable yields and diversification into other cash crops.

In Latin America and in other developing regions, CA is a technology with potential appeal for smallholders. However, adopting CA on a small, possibly non-mechanised, farm involves some different considerations when compared to a large mechanised farm. For example, as smallholders use few purchased inputs, discussions on large increases in herbicide costs may not be relevant. Even if smallholders accept the need for herbicides, they may be unable to finance their purchase. In addition, few smallholders use significant amounts of fertiliser so that a debate over the impact of CA on fertiliser use is largely irrelevant. Ultimately, the availability of credit to assist with CA's increased need for purchased inputs plays an important role. If smallholders hire land preparation equipment, then a switch to CA should be relatively simple as there are no machinery investment implications. Short-run costs would be close to long-run costs when switching to CA.

The majority of smallholders worldwide do land preparation and weeding manually, and adopting CA has its greatest impact on the labour used in these activities. In a comparative analysis of traditional bush fallow systems with no-till and alley cropping in Nigeria, labour savings under the no-till technology were substantial. Whereas alley cropping required from 126 to 151 person-days/ha/year and the bush fallow system needed from 67 to 102 days, the no-till technology required 58 days. These labour inputs amounted to more than 50 percent of total production costs for each technology. However, higher herbicide and equipment costs penalised the no-till technology and it was only preferred under conditions of higher population pressure, which penalises alternative fallow systems.

In studies of smallholders in Latin America, net farm income and returns to labour were much higher under

CA than conventional practice. Table 2 supports this observation for adopters of CA in Paraguay. In judging the attractiveness of CA in smallholder systems in Africa, Latin America and elsewhere, labour savings are a key factor. A further point related to labour is that as the labour savings come at both the land preparation and weeding stages, there are liable to be implications for the gender division of labour.

Table 2. Comparison of conventional and conservation agriculture cropping costs for smallholders at two locations in Paraguay

Crop/cost item (US$ 1998)	Edelira 1/			San Pedro 2/		
	(1) Conventional tillage	(2) Conservation tillage	Ratio (1/2)	(3) Conventional tillage	(4) Conservation tillage	Ratio (3/4) 3/
Farm area (ha)	15.6	15.6	–	6.8	6.8	–
Labour (person-days)	287	240	1.20	164	163	1.01
Net farm income (US$/year)	2 570	4 272	0.60	1 010	2 229	0.45
Return to labour (US$/day)	8.95	17. 80	0.50	6.16	13.67	0.45

1/ average of 3 farms that switched from conventional to a no-till,with green manure crop system.
1/ average of 2 farms that switched from conventional to a no-till with green manure crop system.

In most smallholder systems in Africa, male household members are responsible for land preparation, while female household members are responsible for weeding. Herbicide use may require some adjustment in these responsibilities as male household members usually handle pesticides. Male household members may resist the additional labour demand during the weeding period, so creating a barrier to the adoption of CA. Furthermore, certain conditions can enhance the relative financial attractiveness of CA.

For example, rising land pressure tends to increase the attractiveness of CA relative to bush fallowing. An additional consideration is land quality. Studies of the net returns from mulching, an important component in smallholder CA, suggest that the benefits of this practice increase with the quality of cropland. Successful instances of CA adoption in Latin America have demonstrated the importance of credit as an important enabling factor. This is because of the need to finance specialised planting equipment and herbicides.

Financial Analyses of CA Vs. Conservation Technologies

Most financial analyses of CA concentrate on a comparison with conventional practice, whether this is conventional tillage or bush fallow. However, farmers can often select from a number of alternative conservation practices, in which case CA is just one option of perhaps several. This is especially true for smallholder systems as an absence of prior machinery investments and the small-scale adaptability of many soil and water conservation techniques makes adoption relatively easy in physical and financial terms.

To consider CA's attractiveness in relation to alternative conservation practices to a smallholder, a database of over 130 different analyses of individual soil and water conservation technologies was compiled. The analyses concentrated on Africa and Latin America with all technologies coded according to whether they constituted a CA-related technology or not, as specified by the World Overview of Conservation Approaches and Technologies (WOCAT) technology classification system. Group 1 includes measures aimed primarily at enhancing soil cover and organic matter, while Group 2 technologies are generally linear, cross-slope approaches intended to reduce erosion from wind or runoff. Information about farm-level financial returns was entered in the database for each technology. The results for each of the two technology groups were sorted based on whether technology adoption provided a positive or negative net present value (NPV). Table 3 presents the results of this procedure.

The analysis presented in Table 3 is somewhat crude as many studies employ differing assumptions about project life, discount rates, land opportunity costs, etc. Moreover, the classification of technologies is not precisely consistent with the definition of CA presented earlier.

Table 3. Comparison of financial net present values for conservation agriculture versus other soil and water conservation technologies

Technology	Total number of analyses	Number with positive NPV	Percent with positive NPV
Group 1 Conservation agriculture and related agronomic approaches (e.g. intercropping, contour farming green manure)	40y	34y	85
Group 2 Vegitative, structural and other management improvements (e.g. shelterbelts, terracing, burding agroforestry)	96y	55y	57
Total, all analysesy	136y	86y	65

Nonetheless, the results in Table 3 do indicate that CA and, more broadly, agronomic improvements tend to show higher net returns at the farm level than do other techniques. Arguably, this relative attractiveness of CA is more pronounced than was the case from the comparison of only CA and conventional tillage. Thus, when faced with numerous alternatives to conventional practice, CA and related approaches may offer the best possible returns in many situations. Site-specific factors would determine which individual technology offered the best returns for individual farmers.

In short, the financial evidence in support of CA, a few words of caution are in order. While it is true that CA often conforms to what Pampel and van Es term an 'environmentally profitable practice', this is not always so. Particular location constraints might result in reduced yields, or institutional factors may favour alternative practices. Thus, it is necessary to consider site-specific conditions in determining the financial attractiveness of CA. Even where the financial incentives may appear attractive, a consideration of non-financial factors is required to understand the actual and potential adoption of CA.

Influencing Factors for the Adoption of Conservation Agriculture

A number of studies have sought to identify barriers to adoption beyond the obvious divergence between on-farm costs and wider social benefits under CA. For example:

- Large investment costs may discourage adoption.
- The perceived risk of adopting CA may serve as a barrier.
- Long gestation periods for the benefits of CA to materialise may serve as a barrier to farmers with short-term planning horizons.
- Barriers may be particular to culture and recent history.

In part, the need to consider factors other than net returns reflects farmers' competing objectives in farm management, i.e. profitability versus low investment or minimum subsistence food requirements. Competing technologies may meet individual objectives to varying degrees. In terms of maximising net financial returns, Tables 1-3 suggest that CA can provide better net returns than either conventional practice or other conservation technologies, subject to local site conditions. Table 4 compares various attributes of CA technologies and other soil conservation techniques at the farm level in West Africa. The qualitative analysis applies four criteria representing different smallholder objectives, of which one is financial profitability. While consistent with the net returns analysis in Table 3, the results in Table 4 allow for a much broader evaluation, highlighting assorted shortcomings or advantages of individual technologies that may not be apparent in a financial analysis alone.

Table 4. Factors influencing the attractiveness of conservation agriculture practices at the farm level in West Africa

Soil management techniques	Financial attractiveness (net returns)	Initial effect on yield	Incremental investment	Incremental labour required
Conservation Agriculture				
Mulching	++	+	+	–,+
Ridging	–,++	+	+	––,+
Strip cropping	–,++	–,+	+	+
Alley cropping	–	–	–	––,-
Woody fallow	+,++	+	–– –	––,+
Vegetative and structural				
Vetiver grasslines	–+	– ––	–	–++
Fanya juu bunds	–	–	+	++
Stone faced terraces	–	+	+	–
Tree shelterbelts	-	–	––	+,++

The influences other than net returns shown in Table 4 represent only a small subset of the many non-financial factors thought to influence conservation technology adoption. Farm-level factors vary from farm operation to farm operation and higher level factors are also at work, such as the transmission of information. Furthermore, the variables discussed below, and their broader categories, do not act independently, but rather interact to influence adoption.

Characteristics of Farmer

Since Ryan and Gross first showed that the adoption of agricultural innovations is typically uneven from farmer to farmer, researchers have directed attention to certain characteristics and attributes of farmers in an effort to explain this unevenness. In the case of soil conservation technology adoption, Gould *et al.* emphasize awareness on the part of farm operators to soil erosion or other soil problems as an obvious prerequisite to adoption. Indeed, farmer awareness or perception of soil problems is frequently found to positively correlate with CA adoption. Similarly, the central place of information and knowledge in CA adoption, in terms of being aware of soil problems and potential solutions, should lead the level of education of a farm operator to correlate positively with adoption. Education, be it specific or general, generally correlates positively with the adoption of CA practices, notwithstanding some findings of insignificance or even negative correlation.

Age and/or experience are difficult factors to link to CA adoption, given that studies have shown both a positive and negative correlation. Based on a study of conservation tillage adoption in Wisconsin, Gould *et al.* showed that older and more experienced farmers were more likely than their younger colleagues to recognise soil problems. However, they were less likely than their younger colleagues to address the problems once recognised.

In contrast, several studies have found that income correlates positively with the adoption of soil-erosion control practices. Studies of the adoption of conservation tillage and other CA-type practices have often given significant attention to farm size. Many studies have found that farm size correlates positively with adoption. However, other studies have shown no significant relationship, or even a negative correlation. Hence, the overall impact of farm size on adoption is inconclusive.

Some studies have found that the presence of soil erosion and other soil problems on the farm correlates positively with conservation tillage adoption. However, farmer awareness of and concern for soil problems is probably the more critical factor affecting adoption. Another important farm characteristic is underlying land productivity. In the case of no-till and mulch tillage, Uri shows that in the United States adoption is more likely on farms with low rather than high levels of soil productivity. In addition, a good fit between CA and the farm's production goals encourages adoption.

A more complex factor liable to affect adoption is land tenure. In simple terms, privatising land should lead to better incentives for the adoption of conservation technologies. However, studies of the privatisation of land or titling have not shown that this is necessary to motivate sustainable practices and, in some instances, it has had the opposite effect. As a result, it appears that producers may accept titling because it guarantees land rights, but this does not necessarily bring about changes in their land management. In contrast, there are numerous studies indicating that traditional institutions governing access to land resources in developing regions are flexible in responding to internal and external pressures.

Table 5 summarises the empirical evidence provided by a number of African studies addressing both private title and customary tenure. It shows that the former institutional arrangement does not bestow any advantage over the latter, in terms of investment incentives. Thus, general claims that titling will lead to increased investment in land improvements should be viewed with caution. Without knowledge of the practices associated with CA via some information or communication channel, adoption is improbable. Indeed, studies of innovation adoption and diffusion have long recognised information as a key variable, and its availability is typically found to correlate with adoption. Information becomes especially important as the degree of complexity of the conservation technology increases.

Table 5. The effect of agricultural tenure and perceived tenure security on conservation technology investment in Africa

Tenure type	Country	Impact on Investment decisions
Private title	Ghana	+/X
	Rwanda/Ghana/Kenya	X
	Uganda	+/X
	Somalia	X
Customary Rights	Zimbabwe	+
	Ghana, Kenya	+/X
	Rwanda	+
	Burkina Faso	X
	Niger	+

Information sources that positively influence the adoption of CA-type practices can include: other farmers; media; meetings; and extension officers. However, with respect to this latter source, Agbamu shows that contact alone will not promote adoption if information dissemination is ineffective, inaccurate or inappropriate. Studies have not always shown that the ease of obtaining information correlates with adoption.

Biophysical and Technical Factors

In technical terms, the characteristics and availability of CA technologies are crucial factors in adoption. However, de Harrera and Sain note that availability does not imply individual ownership of the necessary machinery as lease/hire arrangements proliferate. Furthermore, potential adopters must believe that the technology will work. Technical factors interact with biophysical factors, e.g. soil type, rainfall or topography can encourage/facilitate or discourage/limit CA adoption. While some studies have shown that farm operations located within regions of steep slopes and erodible soils have a greater tendency to use CA practices, other studies have found these variables to be insignificant.

Social Factors

CA adoption is seldom strictly a function of individual profit maximisation alone, but also can reflect non-individual or societal interests. More specifically, Lynne argues that farmer decision making usually reflects a compromise between private economic utility and collective utility. Producers often identify this latter interest as 'the right thing to do', at least in those places where stewardship is part of the cultural norm. The argument runs that for many producers the pride associated with stewardship makes up for limits in financial rewards. Examples of such stewardship motives governing land management arrangements include the Landcare movement in Australia.

In contrast, Van Kooten *et al.* modelled the trade-offs between stewardship and net returns on wheat-fallow farms in Saskatchewan, Canada. Their study found that farmers make improvements in agronomic practices to benefit soil quality only under extreme degrees of concern. This result holds despite such practices representing no more than a 5 percent sacrifice in net returns. In addition to stewardship motives, collective action may be necessary to implement CA on a regional basis.

Cooperative arrangements govern numerous activities within village agricultural systems. Although the discussion usually focuses on common property resources, even private land use may overlay with cooperative arrangements governing various aspects of farm management. For example, contour ploughing, stone lines and other structural works require cooperation amongst several or many farmers in order to be effective conservation strategies. Many dimensions of CA fit the cooperative model, including the formation and operation of farmers' groups, dissemination of information, pest control and the purchase of agrochemical inputs.

If CA requires collective action or high levels of social organisation to help it gather momentum, then widespread adoption may be related to a society's social capital. The role of social capital in fostering or retarding the collective action needed in promoting new conservation technology is of growing interest. In the broadest sense, social capital refers to the interconnectedness among individuals in society and considers relationships as a type of asset. Several studies have examined the influence of social capital on technology adoption in either developed or developing countries. For example, kinship, or more exactly 'connectedness to others', can influence the adoption of conservation technology.

3

Conserving Agricultural Ecosystem

Agricultural systems have the potential to improve the environment, to be the custodians of critical aspects of farm genetic diversity needed for future agricultural improvement programmes around the world, and to host vast populations of wild biodiversity. Yet our production systems have not been developed to do so, in recent history. While modern agriculture has brought vast increases in productivity to the world's farming systems, it is widely recognised that much of this may have come at the price of sustainability. The bountiful yields of modern agriculture have been "purchased" with high levels of nonrenewable and often toxic inputs such as fossil fuels, inorganic fertilisers, pesticides and herbicides.

In developing countries, these inputs are often imported and strain the importing country's balance of payments. Often, purchased inputs are used primarily on export crops, and do not improve local food security. We know that conventional agricultural practices displace nature. Instead of letting the ecosystem control itself, control comes from outside. Industrially-produced fertiliser substitutes for relationships between plants and nitrogen-fixing bacteria, overriding rather than working with agroecosystems. Pesticides and insecticides replace equilibrating mechanisms such as pest and insect predators.

But it is equally possible to reverse this approach, to find the ways and means to restore and build on the resilience and strength of the agroecosystem in the struggle to fight pests, diseases or soil deficiencies, or to augment crop yields through pollination. The component of agrobiodiversity which we call "agricultural ecosystem services" is a potent tool to be encouraged and fostered. Agrobiodiversity can be used to re-establish natural balances in farming systems with healthier environments, a more rational use of resources, and a greater dependence on internal rather than purchased controls. And to do this, the primary technique is "biodiversification" of agroecosystems, to evoke self-regulation and sustainability.

Principles for Conservation of Agricultural Ecosystem

Key principles for the conservation of agricultural ecosystem services are:

- It is important that everyone- farmers and policymakers both- understand the concept that agricultural ecosystem services can sustain themselves with proper design.
- Ecosystem services have the potential to reduce both off-site inputs and on- and off-site pollution.
- Promoting identification and taxonomy is necessary.
- Assessment of risks over time, relative dependence, and sustainable livelihoods are critical issues for agricultural biodiversity, and need to be in appropriate balance.

— Policy makers are biased toward large scale plans, whereas much of agrobiodiversity is fine-scaled.

— Costs and benefits of agrobiodiversity goods and services need to be identified.

— Costs and benefits need to be distributed on the basis of careful assessment of possible trade-offs, paying attention to incentives and subsidies, and making them appropriate.

— Creating popular awareness and education is necessary for change

— It is necessary to enhance capacity for adaptation to change

Understanding Agricultural Ecosystem Services

This principle is a key concept behind the notion of ecosystem services, and needs to be understood thoroughly: we often take ecosystem services for granted, as they are provided essentially for "free", so long as we do not abuse the ecosystem. When we lose them, however, we end up paying mightily, for soil fertility amendments, pest control services, and pollination services. It may be helpful to give illustrations of how ecosystem services function, so that we can look at what our policies and plans are ultimately aiming at.

Let us consider some illustrations of each of the three main agricultural ecosystem services: pollination, soil biodiversity, and biodiversity that mitigates pests and diseases.

Natural pest control

An estimated 99 percent of potential crop pests are controlled by natural enemies, including many birds, spiders, parasitic wasps and flies, lady bugs, fungi, viral diseases, and numerous other types of organisms. These natural biological control agents save farmers billions of dollars annually by protecting crops and reducing the need for chemical control. Scientists have tried to then identify what it is, in an agricultural system, that encourages natural biological control agents. A review of 150 published papers on biodiversity effects on pest occurance found that in 53% of the cases plant pest density was decreased in diversified systems, in 18% pest density increased, in 20% a varied response was observed, and no change was observed in 8% of the analysed cases.

The mitigating effect of biodiversity appeared to be caused by herbivore response to diversification rather than by enhanced natural enemy activity. While the exact ecological mechanism that determines the effects of agricultural diversification on pests is still not clear, many applied ecologists have turned their attention to plant diversification in their attempt to increase biodiversity and decrease pest impact.

Soil biodiversity

Soil organisms play a central role in the decomposition of dead organic matter and wastes, and this decomposition process also renders harmless many potential human pathogens. People generate a tremendous amount of waste, including household garbage, industrial waste, crop and forestry residues, and sewage from their own populations and their billions of domesticated animals. A rough approximation of the amount of dead organic matter and waste (mostly agricultural residues) processed each year is 130 billion metric tons, about 30 percent of which is associated with human activities.

Fortunately, there is a wide array of decomposing organisms-ranging from vultures to tiny bacteria-that extract energy from the large, complex organic molecules found in many types of waste. Like assembly-line workers, diverse microbial species process the particular compounds whose chemical bonds they can cleave and pass along to other species the end products of their specialised reactions. Many industrial wastes, including soaps, detergents, pesticides, oil, acids, and paper, are detoxified and decomposed by organisms in natural ecosystems if the concentration of waste does not exceed the system's capacity to transform it.

In agricultural systems, it is important to make efforts to keep a viable component of soil biota, as many of the changes that people impose on soils when they farm are not advantageous to soil organims. It has been well documented, that the conversion of natural vegetation to other land-uses, including agriculture, results in change in the diversity of the soil community. As land conversion and agricultural intensification occur, the planned biodiversity above-ground is reduced (up to the extreme of monocultures) with the intention of increasing the economic efficiency of the system.

This impacts the associated biodiversity of the ecosystem—micro-organisms and invertebrate animals both above and below ground—lowering the biological capacity of the ecosystem for self-regulation and thence leading to further need for substitution of biological functions with agrochemical and petro-energy inputs. We know that soils change under agriculture, but we do not know the critical point at which biological processes cease to dominate, and purchased inputs must compensate. The detection of critical thresholds for functional change is however still a matter of debate. The high biodiversity within many functional groups has been interpreted as conveying a substantial degree of redundancy to the soil biota and led to suggestions of high resilience.

There are a wide range of 'soil bio-technologies' that have the potential to increase and sustain productivity that are currently under-utilised because of the lack of critical evaluation for application to small-scale agriculture. The soil biota may be manipulated by both direct and indirect means. Direct management includes inoculation with species of soil biota, including nitrogen-fixing bacteria, mycorrhizal fungi, control agents for pest and diseases and beneficial macrofauna such as earthworms. Modern molecular research is also increasing the potential for genetic manipulation of some of these organisms prior to inoculation. Indirect management is achieved through manipulation of the living plant and cropping system, organic matter of differing resource quality, other soil amendments and soil tillage.

Agricultural pollination as an ecosystem service

One third of human food is derived from plants pollinated by wild pollinators. Without natural pollination services, yields of important crops would decline precipitously and many wild plant species would become extinct. In the United States alone, the agricultural value of wild, native pollinators—those sustained by natural habitats adjacent to farmlands—is estimated in the billions of dollars per year. As many agricultural landscapes across Europe and North America have become uniform over vast areas, traditional nesting sites for pollinators on farms, such as hedgerows, field margins and other "waste places" have been eliminated. Pollination services provided to crops under modern agriculture have declined precipitously, and there is a recognised need to conserve these services.

Reduction of both Off-site Inputs and on and Off-site Pollution.

In all agro-ecoystems, the cycles of land, air, water, and wastes have become "open": meaning farm productivity depends on inputs acquired from off the farm, and produces outputs that are not recycled but need to be disposed of. This is true of all modern agriculture, but it occurs to a larger degree in industrialised commercial monocultures than in diversified small-scale farming systems dependent on human/animal power and local resources.

The modern farming systems that are replacing traditional practices around the world and achieving high levels of productivity by "controlling" the environment but these modern systems require large amounts of imported energy to accomplish the work usually done by ecological processes in less disturbed systems. Agricultural research and development is at a new turning point now, to find ways to reduce the wasteful resource use patterns of modern agriculture, and eliminate its detrimental impact on the environment by harnessing ecological services.

The search for self-sustaining, low-input, diversified and energy-efficient agricultural systems is now a major concern of some researchers, farmers, and policymakers worldwide. Key strategies for low-input sustainable agriculture are primarily achieved through the overall design of the farming system to promote fundamental ecosystem services. From a management viewpoint, the basic components of a sustainable agroecosystem which will enhance these functions include the following:

— The basic concepts of a low- external input sustainable and diverse agricultural systems must be synthesised into practical alternative systems to suit the specific needs of farming communities in different agro-ecological regions of the world.

— One way of promoting sustainable agriculture is to simply improve farm management. Well-managed farming systems nearly always use less synthetic chemical pesticides, fertilisers and antibiotics per unit of production than comparable, but less well-managed farms. Reduced use of these inputs lowers production costs and lessens agricultures's potential for adverse environmental and health effects without necessarily decreasing, and in some cases increasing, per-acre crop yields and the productivity of livestock management systems.

— Another approach to sustainable agriculture is to aim for a common standard that incorporates all of the above practices; the most viable of these is the organic farming movement. Organic farming is a production system that sustains agricultural production by avoiding or largely excluding synthetic fertilisers and pesticides. Whenever possible, external resources, such as commercially purchased chemicals and fuels are replaced by resources found on or near the farm. Whether certified or not, or wholly organic or not, farming systems that aim for the standards of organic farming are learning to depend on ecosystem services. However, it should also be noted that lower-input farming practices of organic agriculture typically require more information, trained labour, time and management skills per unit of production than conventional farming. Thus, extension services and farmer-to-farmer sharing of information is of major importance.

Identification and taxonomy is essential for supporting agricultural ecosystem services. In many areas of biological diversity, there is a need for greater species identification and taxonomy. But nowhere is the need greater than for those organisms involved in agricultural ecosystem services: pollinating insects, soil fauna, and biodiversity involved in pest regulation. These are the "little things that run the world": insects, for example are far more numerous than any other taxa on earth.

Having the correct identification name gives access to correct and relevant information, a wrong name leads to false and irrelevant information. Taxonomy is that branch of science dedicated to discovering, identifying, naming and classifying organisms (species) and understanding their relationships. It provides an information system based on these relationships and is the foundation of all biological science. It allows us to distinguish individual species from amongst the millions with which we share this planet, and to understand their place, role and functions within living systems.

The state of knowledge of small organisms, such as insects and microbia that figure so prominently in ecosystem services, is still very rudimentary. Yet taxonomists are themselves an endangered species. For example, the following information has been compiled on the number of practising bee taxonomists associated with major institutions by country or region, as of 1999: As can be seen, young taxonomists are not replacing older and retiring specialists. New techniques are desperately needed to make taxonomomic tools that can be used by nonspecialists.

However, there are a number of attempts to make the identification process which is so critical to soil biota, pollination conservation and natural pest control easier for non-specialists to use. A large intergovernmental initiative, BIONET, has been set up for taxonomic capacity building in developing countries, which seeks to link experts and needs around the world, particularly for agriculture- related taxonomic identifications.

Ways of training parataxonomists have been explored for pollinating bees; a simplified key to the genera of bees of North and Central America has been developed in English and Spanish, and three 10-day sessions of the "Bee Course" have been held in the southwestern US, bringing together an international group of pollination biologists and field scientists with experienced bee taxonomists.

Biased Policy Making in Agribiodiversity

As we seek to develop more ecologically-based agricultural development programmes, new understandings and capacities will need to be built among agricultural policy makers, the agricultural research community, and farmers. Modern agricultural research has sought to find technological packages that can be applied across a wide range of heterogeneous agricultural environments and circumstances, usually by making the environment more uniform, with irrigation and agricultural inputs. The challenge now is to find more site-specific ways of managing agricultural lands.

Cropping systems and techniques tailored to specific agroecosystems result in a more fine-grained agriculture, based on appropriate traditional and improved genetic varieties and local input and techniques, with each combination fitting a particular ecological, social and economic niche. The participation of farmers, who have this intimate knowledge of the lands they manage, becomes of even greater importance.

For the key functional groups providing agricultural ecosystem services (pollinators, soil biota, natural enemies of agricultural pests) we do know of a number of best practices for promoting their persistence in an agricultural landscape. How to translate this into agricultural policy will need the concerted attention of national agricultural planners and policy makers, in dialogue with biodiversity planners.

We know that following the set of best practices below will promote the ecosystem services to be provided by pollinators, nutrient recyclers, and natural enemies of crop pests. Few farmers will have the time or resources to address each "Best Practice" in turn, but many of them are subsumed under general land conservation activities. With a small investment in further public education, farmers who are already inclined toward good land stewardship may learn to watch out for the small resource areas or special considerations that beneficial farm biota require. With more emphasis on farmers' learning structures, farmers can learn from each other.

Best practices

Diversity of land use can be attained by a variety of methods. They include crop rotations, adapting choice of crops and cultural methods to soil and soil moisture conditions, interplanting crops in a mixture (polyculture), or planting crops among useful trees either conserved in land preparation or deliberately planted (agroforestry). There are also various mixed farming systems in which arable, trees and livestock are integrated. The values of maintaining landscape level diversity in any of these ways include maximising the use of land, managing pests and diseases, sustaining habitats for pollinators and other useful biota, and enhancing biodiversity in the soil, as well as sustaining both floral and faunaldiversity. It should be kept in mind that it is much more common to adopt different management methods on different micro-environments on small farms.

Assessment of Risks over Time and Relative Dependence

The objectives of ecosystem management are the optimisation of sustainability, the minimisation of risks and the maximisation of ecosystem services. National planners must carefully address the issue of minimising risks, for the sustainable livelihoods of their citizens. In relation to agriculture, many of the inherent risks relate to decisions on appropriate agricultural yields, and how to obtain those yields. A national agricultural policy on sustainable agriculture includes, whether implicitly or explicitly, the yield expectations of the country; a farmer's approach to production also includes a yield expectation. If it is expected that yields must be high, a high level of risk must be expected. Other systems tolerate lower yields and lower risks.

This is a matter of societal choice, as well as ecological factors. One aspect of high risks is a heavy dependence on outside sources of material or technology. A transition to more sustainable agriculture has added benefits, in terms of minimising risks, in that the farming system becomes more "closed", using internal inputs and ecosystem services rather than outside, purchased inputs. The model of Cuba, as it made a rapid nationwide conversion to organic agriculture, is a good object lesson not only in the efforts needed, but also the benefits possible by such a transition.

Costs and Benefits of Agrobiodiversity Goods and Services

To get these services properly entered into national economic accounting, we need to have highly realistic, well documented analyses of costs and benefits. Thus, the essential best practices with respect to each agricultural ecosystem service, are:

Even in the absence of detailed valuations of agricultural ecosystem services, there are many initivatives underway to develop markets for organic agriculture, which includes the promotion of most agricultural ecosystem services. GTZ, for example, supports several programmes that seek to build such markets. One of the these, Protrade, encourages public-private-partnerships. Specialising in trade and business promotion, the program offers assistance in sector-related marketing, product and production consulting in more than 90 countries, promotion in Germany and the EU, trade fair assistance and a comprehensive information service.

Protrade included an organic products sector in 1993 in reaction to the growing demand for biological cultivation of products and the strong interest of many third world countries in organic agriculture and farming. The main emphasis of the work in the organic products sector is on developing new trade contacts, consulting in the areas of organic farming, certification, product development, management and quality assurance, as well as offering support for participation and international specialist trade fairs. Fifteen countries are currently in the consulting programme: the Dominican Republic, Ecuador, Haiti, Honduras, Kenya, Madagascar, Mexico, Nicaragua, Peru, Russia, Senegal, Zimbabwe, Uruguay and Tanzania.

One key element of building markets, and assuring that farmers receive sufficient incentives for conserving agricultural biodiversity, is to develop certification systems that guarantee organic products for consumers, and higher prices for producers. TransFair is a seal offered by the labelling initiative "Transfair International" to traders who buy from registered cooperatives in developing countries and abide by fair trade criteria.

Products covered by the TransFair seal include coffee, honey, cocoa, sugar and tea. Several other initiatives like TransFair are all grouped in the Fair Trade Federation. They, and other important actors in the field of labelling organic products can be found on the websites given in Tools, below.

Along with building proper incentive measures, it is important to eliminate pervese subsidies, of which there are many in the agriculture sector. There is a growing recognition that currently most if not all policy measures used to support agriculture act as powerful disincentives against sustainable agriculture, especially with respect to subsidies for agricultural inputs. Countries have adopted these policies because they believe that the economic benefits outweigh public expenditure. But in fact, recent research shows that the use of chemical pesticides has been promoted and subsidised by governments even though there has been little information on the net benefits of a dependency on pesticide usage compared to other crop protection strategies (GTZ Pesticide Policy Project, no date).

Agenda 21 of the United Nations Conference on Environment and Development in 1992 demands the implementation of integrated pest management (IPM) as an alternative to dependency on unilateral use of agricultural chemicals. This concept stresses the use of local knowledge and aims at improving the decision making capacity of farmers instead of disseminating fixed packages of external technology.

But the paradigm switch that is indispensable for the dissemination of IPM will not be obtained without a change in the general institutional framework of economic incentives that govern crop protection. In many instances, national governments hesitate to implement the necessary political changes. They lack adequate concepts for policy analysis in a sub-sector that is dominated by specialists in natural science and toxicology.

Agricultural interest groups fear that policy reform measures aimed at a reduction of chemical pesticides will threaten their competitiveness in global agricultural product markets. The economic justification of pesticide use in world agriculture and horticulture is often doubtful. Negative side effects occur in many instances but are not accounted for as costs (see box, Negative Side Effects of Pesticide Use). For example, farmers do not consider occupational health costs in their decision to use pesticides. The costs of the damage of the natural environment are born by the society as a whole and only perceived in the long run. They are not embodied in the private costs of pesticide application. Therefore, actual pesticide use often exceeds the social optimum.

There are a number of new resources and initiatives being launched to assist countries to better examine existing policy decisions, and undertake a reform of crop protection policy instruments within government, and private sector approaches.

Enhancing Capacity for Adaptation to Change

If we are to make sweeping changes in the way our food is produced, and institute a greater reliance on ecosystem services, it should be recognised that under any "change management" there are transaction costs.

The process of conversion from a high-input conventional management to a low-input (or low-external input) management is a transitional process which can have considerable costs for farmers. There are four identified phases in making this transition:

1. Input withdrawal: progressive chemical withdrawal.

2. Efficient input use: rationalisation of agrochemical use through integrated pest management (IPM) and integrated nutrient management.
3. Input substitution: using alternative, low-energy inputs
4. System redesign: redesign of diversified farming systems with an optimal crop/ animal assemblage which encourages synergies so the system can sponsor its own fertility, natural pest regulation and crop productivity.

During the four phases, management should be guided in order to ensure the following processes:

1. Increasing biodiversity both in the soil and above ground
2. Increasing biomass production and soil organic matter content.
3. Decreasing levels of pesticide residues and nutrient and water loss.
4. Establishment of functional relationships between the various farm components.
5. Optimal planning of crop sequences and combinations and efficient use of locally available resources.

As well, there are a number of constraints that the "change management" regime must contend with. First labour inputs are substantially greater for organic technology than for conventional production; productivity may average 22 percent to 95 percent less than under-conventional production. Another constraint is adequate quantitites of organic fertiliser like manure. As livestock production is becoming more and more concentrated in feedlots, whole regions are finding that a shortage of manure on farm is complemented by toxic concentrations of animal waste on feedlots, often near urban areas. This poses the potential for developing sound regional ecosystem interdependencies, but the systems to share such resources in a rational manner need to be developed.

Capacity is another large constraint for organic farmers. A study by Blobaum of farmers in the United States noted that several capacity- and information-related obstacles discourage conventional farmers from adopting organic methods. Organic farmers perceive the lack of access to reliable organic farming information as a serious barrier to conversion. Most rely primarily on information from other organic farmers and from such non-traditional sources as books and magazines, representatives of organic fertiliser companies, and workshops and conferences. Organic farmers have a strong interest in research on many problems, and most farmers would adopt new practices if more research-substantiated information were available.

Blobaum also found that organic farmers who use special markets are dissatisfied with problems such as small orders, long delays in getting paid, inadequate returns for cleaning and bagging grains, confusing certification standards, difficulty in contacting buyers, and the expense of maintaining special on-farm storage areas. Credit discrimination is seen as a potential problem by a sizeable number of organic farmers. The long-term economic benefits of organic agriculture may not be evident to a farmer faced with having to make payments on annual production loans. Many conventional farmers are greatly in debt, and their debt constrains the shift to more sustainable methods.

Encouragingly, recent economic evaluations suggest that profits from organic farms can exceed those of conventional farms. Eventually, farmers should be able to reap economic benefits from practising a more sustainable production system. But farmers' inability to overcome the multiple constraints, as listed above, may impede the transition to more sustainable agriculture. The costs of making the transition should not, and cannot, be borne by farmers alone. Governments and non-governmental organisation should seek ways of helping the agriculture sector adapt to changes in ways that benefit all of society.

Governments should appreciate that changes need to be phased and strategically designed. In the case of Denmark, which aims to become a pesticide-free country within five years, considerable work has gone into developing a phased, methodical transition to alternative technologies. The research and logistical development which will be necessary to support this transition come from investment on the part of the Danish government. Non-governmental organisations can also play important roles in making transitions, particularly when they assist in developing "redesign strategies".

Altieri found that in Latin America, NGOs working with communities and applying agro-ecological methods have shown that transitions to organic production need not be the perogative of developed countries, or naturally productive areas. Even within a policy environment that has not been conducive to sustainable agriculture with a strong political primacy of urban social groups, heavy dependence on industrial production, the absence of effective land distributions, subsidies for fossil-fuel based agricultural inputs, and the limited access of peasants to political and economic resources, NGOs working with farmer groups have been able to show that low-external input practices can be economically, socially, culturally, and ecologically appropriate for those farmers who have not benefited from conventional agriculture, or farm in marginal areas.

Creating Popular Awareness and Education

Ecosystem services may be important, but few people know about them or understand them. Messages on ecosystem services are needed, adhering to the following best practices:

1. Agriculture depends on the wider environment.
2. Agriculture derives more from environment than technology can substitute.
3. Sustainable agriculture benefits from biodiversity, such as water quality improvement and pollution reduction.
4. Biodiversity benefits can feed back into increased agriculture production, e,g., pollination, or natural pest control can contribute to yield increases.
5. Conversely, loss of pollinators can lead less production and economic benefits
6. These benefits can be quantified.
7. Agricultural practices that maintain soil organic matter also maintain soil diversity.
8. Sustainable economic development is dependent on agrobiodiversity.

A number of creative means need to be undertaken to convey these messages to the public.

In the case of the pollination research carried out in Kenya, research findings were given back to the community and the general public in the form of a colourful poster, distributed through a popular journal. Other innovative public education programmes in relation to soil biodiversity have been started in Australia and Canada.

Public awareness programmes addressing ecosystem services will need to draw on the resources of scientists, and help scientists to work together and with other stakeholders. The Brazilian Government Proposal to the Convention on Biological Diversity Subsidiary Body on Scientific and Technological Advice identified institutional and educational factors as the main constraints to effective use of soil biodiversity. These included: low institutional capacity, lack of integration between different groups working on the topics, insufficient information exchange, and lack of public education to appreciate the value of the soil biodiversity. The following goals and strategies were proposed in alleviation:

1. Establishment of a network of laboratories, scientific collections and technical centres.
2. Programmes for education of specialists
3. Definition of standard sampling protocols
4. Definition of indicators of soil quality
5. Develop models to measure the economic value of the biodiversity of microorganisms and creation of fiscal incentives.
6. Establishment of specialised discussion groups of researchers
7. Establish thematic networks on soil and micro-organisms biodiversity.
8. Education targeting the appreciation of the value of the biodiversity of microorganisms and their sustainable use and development. These actions could act as a model for other countries.

4

Field Preparation and Planting

Site preparation is used to create favourable growing conditions for hardwood seeds and seedlings, and to facilitate tree planting operations. Site preparation includes controlling undesirable vegetation; improving soil structure, drainage and fertility; and reducing logging slash.

Site preparation is the creation of a favourable growing environment for tree seeds or seedlings. The biggest obstacle facing seedling establishment is competition from other vegetation. Effective site preparation will reduce competing vegetation and create a sufficient number of suitable growing sites without causing excessive soil disturbance.

Mechanical Site Preparation

Mechanical site preparation typically disturbs the soil and reduces competing vegetation by plowing, disking, raking, chopping, scalping, and trenching, among other approaches. In some cases, mechanical site preparation can be valuable, as exposing and disturbing mineral soil can have the added benefits of increasing root zone temperatures, aerating the soil, and improving drainage. However, when mechanical site preparation is done incorrectly it can result in soil erosion, compaction and rutting. Also consider the potential competing vegetation that may develop from dormant seeds after the soil is exposed. A combination of both mechanical and chemical techniques can be used for added control.

Chemical Site Preparation

Chemical site preparation can be an effective method to control vegetation and increase the amount of sunlight and water available for plant growth. Chemical methods may involve simple equipment, can be less expensive and provide longer control than mechanical site preparation. However, chemical effectiveness depends on the appropriate herbicide selection, the timing of application, application rate and weather conditions. Herbicide applications may need to be repeated for several years to ensure stand establishment. All herbicides must be applied in accordance with label recommendations and their registered use.

Prescribed Burning

Prescribed Burning, or controlled ground fires, can be an effective and inexpensive means of removing or reducing vegetation and preparing a suitable seedbed. Burning can also improve soil nutrient levels and ectomycorrhizal development. Prescribed burning, however, can reduce the effectiveness pre-emergent herbicides and can increase solar heating at ground line leading to seedling mortality. The use of fire as a vegetation management technique is very appealing to many small landowners because it appears 'natural'- but it can be dangerous. Effective and safe use of prescribed fire requires appropriate equipment and training.

Cover Crops

Cover Crops are appropriate for afforestation sites where they are grown to prevent invasion by noxious weeds, non-native invasive species, or other competing vegetation. Cover crops can also control soil erosion, improve soil condition and increase water-holding capacity. When selecting a cover crop, choose a species that will accomplish the site preparation objectives but not adversely impact tree growth. Legumes are sometimes selected as cover crops because they can enhance soil nitrogen. Small grain crops, such as winter wheat and rye, can inhibit weed growth and add organic matter to soils while providing limited competition for tree seedlings. Winter wheat can be spring seeded to produce a less vigorous but effective cover crop.

Former Agricultural Fields

Former agriculture fields present a unique set of site preparation challenges. Fields that were in row crops such as corn or soybeans the previous year generally require a pre-emergent herbicide after planting to control germination of stored weed seed. Cover crops may also be used to control invasive weeds. Alfalfa, clover, or some perennial grasses provide fierce competition for tree seedlings and seeds. Alfalfa and sod are easiest to control during the year prior to planting with an early fall application of herbicide when the plants are still actively growing. Alternatively, rotation into a row crop or other desirable cover crop, followed by planting of seedlings, has been especially effective for hardwood plantings on heavy soils.

Preparation and Planting of Landscape Plants

Plants are important to the microclimate because they

absorb heat from the air during the transpiration process (release of water in the form of vapor), provide shade that reduces solar radiation and reflection, lower wind speed, disperse fog, and influence snow deposition. Businesses have noted that attractively landscaped buildings result in above average labour productivity, lower absenteeism and easier worker recruitment. Studies have indicated people have a basic desire for contact with plants. Satisfying this need can have a strong positive influence on human behaviour. Hospital rehabilitation therapists are aware of this fact and increasingly include gardening in their programs.

Landscape plantings are important features of private and public properties and must be selected, planted, and maintained with skill. The emphasis of this article will be on the planting process with only a brief review of selection and maintenance.

Plant Selection

The first step in selecting plants is based on specific needs identified in the landscape plan. A plan is important to maximise the landscape investment and meet the needs of climate control, noise reduction, privacy screening, or to satisfy other objectives. The plan should address the total property, but may only include the front of the home one year and another area later. The selection process should consider factors such as ultimate size, adaptability to local soil conditions, exposure to sun, shade, or wind, hardiness (ability to withstand cold, drought, etc.) and maintenance requirements. Consider as well, branching habit, plant texture, foliage colour, flowers, and fruit during various times of the year.

Landscape plants are available in several common forms: (1) balled and burlapped (B & B), (2) bare-root, (3) packaged, (4) container grown, (5) field potted, and (6) flats or market-paks.

Balled and Burlapped

Many trees and shrubs are commonly transplanted with a ball of soil wrapped in burlap. Field grown evergreens and most trees with trunks more than two inches in diameter are available B & B.

Bare-root

Dormant plants were historically harvested bare-root (without soil) and offered to the public in this form. Nurseries harvested shrubs and smaller trees in this way and overwintered the bare-root plants in refrigerated storage. Many garden centers, however, do not have facilities to properly control the storage environment during the spring sales season, and therefore, this method of offering plants is being replaced by other practices. Many mail-order suppliers still provide plants in this form.

Packaged

Woody plants, particularly roses and some perennials such as peonies, are sold bare-root with either the root system or entire plant packaged in a cardboard or plastic wrap. Prior to packaging, the root system is wrapped with a moisture-holding medium or packing material. For the highest degree of success, plant bare-root material when they are in the dormant state or in the very early stages of growth.

Container Grown

This has become the primary form in which plants are available in the garden center. They are easy to maintain in the retail store and after purchase, and can be planted anytime of year unless the ground is frozen. Trees are now available in large sized containers from 10 to 25 gallons or more. Many garden centers will accept the return of plastic containers and reuse or recycle them.

Field Potted

This technique refers to plants that have been grown in the field and when harvested, placed in a container (often a fibre pot) rather than wrapped in burlap. Typically, narrowleaf evergreens are produced in this fashion. The plants are dug and containerised in autumn, stored under a polyethylene covering in winter and delivered to retail outlets in spring.

Flats or Market-paks

Ground covers, perennials, and annuals are sold in these containers. There are many different styles and forms available and nearly all are made of plastic which can be reused many times.

Planting Season

If the soil is moist and kept sufficiently watered, planting can be done anytime the soil is not frozen. B & B or container-grown plants can be planted anytime during the growing season. Dormant, bare-root stock is best planted in early spring or late autumn, if available. Some plant species actually survive and grow better if planted in spring. Examples include beech, sassafras, sweetgum, sourwood, walnut, white oak, or others that have few, but large roots. Most plants that have many small fibrous roots can be transplanted with a high degree of success throughout the year.

Site Preparation

Planting depth and a knowledge of soil texture (size of soil particles), soil structure (arrangement of soil particles), and the capacity of the soil to drain are critical. The depth of planting is dependent on soil type or texture (size of soil particles, e.g., sand [largest], silt [next largest] and clay [smallest]). In sandy loam soils that drain well, plants should be positioned in the planting hole at the level they were originally grown in the nursery. The original depth can be identified by a soil line on the stem or trunk of plants or by the top of the soil of B & B or container-grown plants.

Soils that remain wet for long periods of time after a rain or irrigation can be managed to prevent problems. One option is to create planting beds on top of the existing soil. This could involve using landscape timbers to build up an area and filling it in with new topsoil or packaged planting medium. Another option is to install drain tile in the planting hole. The drain tile works more effectively if it is situated on a slight slope so water can drain away from the planting or bed area.

The width of the planting hole should be at least 2 or 3 times the diameter of the root ball. Recent research results have indicated feeder roots of trees and shrubs remain close to the soil surface, usually in the top 6 to 9 inches. This suggests that the major area of soil preparation should be the upper 12 inches, and the wider the area prepared for planting, the better.

If the sides of the planting hole have become compacted or glazed in the digging process, which is quite common in machine digging, use a shovel or spade to scarify or roughen the smooth surface. This provides for better air and moisture movement as well as root penetration.

Planting

Remove any dead, diseased, broken, or twisted roots before planting by pruning to healthy tissue. Roots matted at the bottom or circling around the root ball of container-grown plants should be cut and removed or straightened. Some landscapers cut the root mass of container plants vertically on opposite sides at least half the distance to the main stem to decrease the chance that hidden circling roots will girdle ("strangle") at some time later. In contrast, disturb the root ball of B & B plants as little as possible.

Slowly lower the root ball into the planting hole to prevent breaking or cracking of the soil around the roots of B & B plants. After placing in the hole, the burlap of B & B plants should be loosened and rolled down the sides of the soil ball. No burlap should be exposed above the soil surface in order to prevent wicking and drying which interferes with proper moisture flow between the soil ball and the backfill. Cut and remove all twine, cords, wire, or strapping material, especially around the trunk or stems. Also, remove all labels attached with wire or nylon twine that can eventually girdle stem tissue.

Position the soil ball in the hole with the lowest branches of the plant facing areas of least people and pet activity, and higher branches facing areas requiring greater clearance. This reduces subsequent pruning. Orient the side of the plant with the most branches in such a way they are away from the afternoon sun. In so doing, the less developed side of the plant will be favoured with more light.

The Backfill

If the soil removed from the planting hole is a well-drained humousy loam, there should be little need for amendments. However, most soils in the urban landscape are silt to clay base, have often been disturbed by construction of the home, and, therefore, are less than ideal for plant growth. With these soils, organic amendments are beneficial. Most organic matter amendments will open poorly drained, compacted soil and improve aeration, water infiltration, and percolation. In sandy loam soils, these same materials will increase water-holding capacity.

Certain organic products influence soil pH. Sphagnum peat moss and pine bark chips lower the pH of most soils, while most composts, including leaf mold, will increase soil pH. Having the soil tested prior to planting will help determine the need to adjust pH or mineral element levels. New root growth can be enhanced by adding some form of organic matter to the soil being used to fill the planting hole. All subsoil from digging a planting hole should be discarded, especially if blue, gray, or yellow in colour. Mix the soil amendments thoroughly with the top soil. Work the mixture gently around the root system and firm lightly. Heavy packing is not necessary and may be harmful. Water thoroughly, at least 2 to 3 times in a 60-minute period to make certain the entire growing mix is well moistened.

Leave a saucer-shaped depression around the plant inside the edge of the soil ball to facilitate subsequent watering. Water all container and B & B plants to thoroughly wet the root ball area prior to planting. Small amounts of fertiliser may be added to the backfill when planting. Since phosphorus moves slowly in the soil, it is important to incorporate it, if called for by a soil test, at the time of planting. This can be done by adding 5-10-5, 4-12-4, 0-20-0, 0-45-0, or similar high phosphorus ratio fertilisers to the backfill. Use half a pint of 5-10-5, 4-12-4, or one-fourth pint of 0-20-0 or 0-45-0 per bushel of soil.

Staking

Staking trees after planting may be necessary until new root development is sufficient for the plant to stand alone. Staking may also be needed to protect young trees from vehicles, equipment, animals, and vandalism. Trees with trunks up to two inches in diameter can be supported with a single stake 2 x 2 inches x 8 feet driven firmly into the ground close to the trunk. Attach the tree to the stake with plastic ties designed for the purpose, or with a wire run through a piece of garden hose. The garden hose will protect the trunk from injury by the wire as the tree grows in diameter. Do not fasten a stake to a tree with wire alone. Trees with trunks larger than two inches in diameter may require 2 or 3 stakes or guy wires anchored into the soil.

Maintenance in Planting

Watering

Natural rainfall is usually not adequate to provide the moisture needs of newly planted landscape plants. The limited root system of bare root, container grown, or B & B plants makes them highly susceptible to dry weather conditions. Pay particularly close attention to plantings of container plants. The potting media used for container plants dries out much more rapidly than soil. Determine the amount of soil moisture by sampling the soil with a narrow trowel or soil probe at a depth of 6 to 8 inches. Squeese the soil. If it holds together and is not sticky, soil moisture is adequate. If the soil does not stick together, it is too dry and water is necessary. At each watering, wet the soil thoroughly to the base of the root system. Apply a minimum of one inch of water per square foot per week between rainfall and irrigation. If the temperature is high or the soil is very dry, apply more than one inch per week.

Be cautious not to overwater or the amount of oxygen in the soil will be lowered to a level that will damage roots. Make certain the timing patterns of lawn watering systems are not overlapping into plant beds and too much water is being applied to landscape plants. Monitoring the soil for moisture with a soil probe or trowel, as mentioned above, is recommended.

Mulching

Apply a mulch around newly planted landscape plants to conserve moisture in summer, reduce weed growth, stabilise soil temperature, and improve the overall appearance of the site. Apply two inches of one of the bark products, sphagnum peat moss, compost, or any number of other products readily available for the purpose. Avoid applying too much mulch, as it could significantly reduce drying and lead to excess wetness in clay loam or poorly drained soils. Reapply or replace mulch annually in the autumn, late spring, or both, depending on rate of decomposition. Never allow the mulch to exceed 3 inches in depth.

Fertilising

The key to plant survival and growth is proper moisture management; however, most plants require an annual application of nitrogen, if not a complete fertiliser. Each autumn after the first hard freeze, apply a fertiliser with a 3-1-2, 4-1-2, 3-1-1, or 4-1-1 ratio of nitrogen, phosphorus, and potassium (N-P-K). Apply at a rate of 3 pounds of actual nitrogen per 1,000 square feet annually to maintain healthy plants. For example, if using an 18% nitrogen fertiliser such as an 18-6-12, applying 3 pounds nitrogen to 1,000 square feet would require the use of approximately 17 pounds of product. One half the above amount should be applied to broadleaf evergreens, dwarf conifers and rock garden plants. Apply the fertiliser when the soil is moist and water the nutrients into the soil. Slow-release fertilisers will yield a longer period of release and are safer to use than quick-release farm grade fertilisers.

Pest control

If pest-resistant plants were selected initially, the need for pest control should be minimal. However, weather conditions dictate the degree of many pest infestations and treatment with pesticides may be required from time to time. When insects or diseases are first observed, take a sample of the insect or affected plant part to a local garden center and ask for a diagnosis. Failure to control pests in the early stages can lead to greater problems later in the season.

Pruning

The selection of the proper plants for the site will reduce the need for much pruning over time. However, branches will occasionally outgrow the intended space or plants may be damaged (storms may break up plants, etc.). Pruning of most plants is best done in late winter or early spring, although light pruning can be done anytime. Annual pruning is done to maintain plant shape, or to remove dead, broken, weak, crossing, or low-lying limbs. Removing branches while still less than an inch in diameter promotes rapid healing of wounds.

For an excellent landscape program, select high quality plants and have a knowledge of their growing requirements. Plant at the proper depth based on soil conditions, prepare a large enough planting hole, use a prepared backfill, and water thoroughly. Remove all twine and wires and maintain plants with water, mulch, fertilizer, pruning, and pest control.

5

Crops and Cropping Systems

Domestication of plants is done in order to increase yield, improve disease resistance and drought tolerance, ease harvest and to improve the taste and nutritional value and many other characteristics. Centuries of careful selection and breeding have had enormous effects on the characteristics of crop plants. Plant breeders use greenhouses and other techniques to get as many as three generations of plants per year so that they can make improvements all the more quickly.

Plant selection and breeding in the 1920s and '30s improved pasture (grasses and clover) in New Zealand. Extensive radiation mutagenesis efforts (i.e. primitive genetic engineering) during the 1950s produced the modern commercial varieties of grains such as wheat, corn and barley. For example, average yields of corn (maize) in the USA have increased from around 2.5 tons per hectare (40 bushels per acre) in 1900 to about 9.4 t/ha (150 bushels per acre) in 2001, primarily due to improvements in genetics.

Similarly, worldwide average wheat yields have increased from less than 1 t/ha in 1900 to more than 2.5 t/ha in 1990. South American average wheat yields are around 2 t/ha, African under 1 t/ha, Egypt and Arabia up to 3.5 to 4 t/ha with irrigation. In contrast, the average wheat yield in countries such as France is over 8 t/ha. Higher yields are due to improvements in genetics, as well as use of intensive farming techniques (use of fertilizers, chemical pest control, growth control to avoid lodging).

In industrialized agriculture, crop "improvement" has often reduced nutritional and other qualities of food plants to serve the interests of producers. After mechanical tomato-harvesters were developed in the early 1960s, agricultural scientists bred tomatoes that were harder and less nutritious. In fact, a major longitudinal study of nutrient levels in numerous vegetables showed significant declines in the last 50 years; garden vegetables in the U.S. today contain on average 38 percent less vitamin B2 and 15 percent less vitamin C.

Very recently, genetic engineering has begun to be employed in some parts of the world to speed up the selection and breeding process. The most widely used modification is a herbicide resistance gene that allows plants to tolerate exposure to glyphosate, which is used to control weeds in the crop. A less frequently used but more controversial modification causes the plant to produce a toxin to reduce damage from insects.

There are specialty producers who raise less common types of livestock or plants. Aquaculture, the farming of fish, shrimp, and algae, is closely associated with agriculture. Apiculture, the culture of bees, traditionally for honey—increasingly for crop pollination.

Strategies of Cropping System

Planting and Seed

Seed quality is a key to good yields. For most crops, use

certified seed or seed of a quality comparable to certified seed. Home-grown seed can save money, but you should know the germination so you can get a proper seeding rate. Never use seed from hybrid crops for next year's planting (corn or grain sorghum).

Use New Varieties

Soybean varieties have been improving at about 1/3 bushel per year. Using a 10-year-old variety means you may end up with about 3 less. Corn varieties are improving at about 1 bushel per year. Use current variety test information to help make your decision.

Seed Treatment

Wheat seed treatment is generally beneficial. Soybeans may not need seed treatment in non-stress conditions. Seed treatment can improve some low germination seed. Narrow rows usually increase yields for nearly all crops. (Table 1) Consider narrowing existing equipment to the limit and, as needs arise, purchase equipment for narrow rows. Soybeans planted in narrow rows (10 inches or less) have increased yields an average of 4 over 30-inch rows. Narrow rows show the greatest advantage in late-planted crops such as double crop soybeans or grain sorghum. You can get narrow rows by doubling back with wide-row equipment.

Planting date is extremely important in achieving yield potential. Nobody can plant every acre on that best planting date. But strive to have at least half of your crop planted by that date. The only excuse for late planting should be the weather. Plan ahead to get your planter in field condition before that first planting day. Check and adjust your planter often. Drive as slowly as possible to ensure good seed placement and proper seeding rate while still getting the planting done on time.

Table 1. Average soybean yield in to drilled seeding at several Missouri locations

Year	Location	30-inch rows	10-inch rows	Response
Year 1	Novelty	41.8	42.8	1.0
	Marshall	43.3	44.6	1.3
	Columbia	45.8	51.0	5.2
	Portageville	31.2	34.4	3.2
Year 2	Marshall	47.5	52.9	5.4
	Columbia	47.7	51.7	4.0
	Portageville	39.7	43.0	3.3

Tillage costs money in fuel, labour and wear and tear on equipment. Reducing the number of tillage operations and tillage depth will reduce these costs. Work the soil only until you achieve that seedbed that will give satisfactory emergence. Added tillage may also increase soil erosion potential. Evaluate tillage by categorizing necessary and traditional tillage.

Herbicides and Weed Control

Too often, farmers select herbicides for broad spectrum control when less expensive alternatives may control the weeds present in your crops. Comparing prices of herbicides in relation to their recommended application rates may result in considerable cost savings. Several herbicides are manufactured by more than one company. Shop for the least expensive brand of the same herbicide.

Band application of some herbicides over the row can reduce costs by as much as two-thirds. Some herbicides do not require incorporation and can be effectively used in a band over the row for pre-emergence or post-emergence application. You can cut costs by cutting down on the amount of herbicides used per acre and by reducing tillage.

Use a rotary hoe to control small weeds early and a cultivator to control major weeds as the crop grows. This can result in effective weed control and can cut costs.

Table 2. Nitrogen added to the soil by a legume crop (optimum)

Legume crop	Nitrogen added per acre
Alfalfa	
80-100 percent stand	120-140 pounds
50-80 percent	40-60 pounds
Less than 50 percent	0-20 pounds
Sweet clover (Green manure)	100-120 pounds
Red clover (Pure stand)	40-60 pounds
Soybeans	15-60 pounds

For weeds in the rows, post-directed or rope wick application may give the added control you need at a reduced cost. Take care in selecting fields for this purpose. Be certain that under almost any condition, timely cultivation and herbicide application can be accomplished.

Fertilizers and Soil Fertility

Get a soil test before applying fertilizer. Fields high in phosphorus or potassium can go without those nutrients for a year or more without reducing yields. If you have some fields with high fertility and some with low, fertilize only those with low fertility. This practice will give a better return than fertilizing all fields at a reduced rate. When cash flow is good, fertilize to high levels. This will allow you to reduce fertilizer rates when cash flow is poor.

Sound fertility management is just as important during tight times as it is when cash flow is adequate. Avoid the temptation to buy unproven products; stick with the proven and recommended practices. New miracle products often result in no increased yields.

Stick to the major nutrients during these economic times. As yet, many micro nutrients are not recommended by University agronomists. The probability of response to micro nutrients is much less than with N, P or K. It is easy to increase a fertilizer bill $10 or more per acre with micro nutrients. Use only those nutrients with a proven response on your soil.

Consider band application of fertilizer rather than broadcasting. Band application has long been proven to be more efficient at lower fertilizer application rates. Banding can be accomplished at planting by placing fertilizer near the seed or by using new innovations, such as knifed placement or dribble applications.

Ensuring effective use of applied fertilizer is always important. The key is incorporation of the fertilizer into the rooting zone. Potassium and phosphorus applied at planting without incorporation often result in little or no response. Incorporate these nutrients at fall or spring tillage.

Remember to reduce nitrogen rates appropriately when legumes or manure are used in crop rotations. (Table 2.) Soybeans can add 30 to 50 pounds of N to a subsequent wheat or corn crop. Manure commonly contains appreciable amounts of N, P and K. To effectively reduce the fertilizer needed, determine the nutrient content of manure by lab analysis. Fresh animal manure often contains 10 to 15 pounds of nitrogen (N), 5 to 10 pounds of phosphate (P_2O_5), and 10 to 15 pounds of potash (K_2O) per ton.

Crop Growth

Monitor fields periodically for insect, disease, nutrient or other associated problems while the crop is actively growing. Don't use fungicides or insecticides until the economic threshold of crop damage has been reached. Consult an agronomist on field problems to help you make those costly decisions. One less spraying could save several dollars. According to research, twenty percent leaf loss in soybeans does not normally reduce yield potential.

Monitoring crop growth this year can help you plan to correct next year's problems. Fertility problems are often difficult to overcome during the current cropping season but, through plant analysis, you can make plans for next year. Avoid unnecessary trips to fields. During active growth, weekly, thorough inspection is necessary. But as growth tapers off, growth problems are less likely.

Crop Rotations

Crop rotations have consistently shown yield increase (roughly 10 percent) over continuous mono-cropping. Plan crop rotations as best fit your land and potential crop prices. Consider planting as many soybeans as possible for short-run cash flow improvement. Production costs for soybeans are generally lower than most other row crops.

Plant suitable acres to double crop soybeans or grain sorghum. With a minimum cost, you can get some return from continued use of crop acres through double cropping. Avoid double cropping on unsuitable land or during dry conditions. Don't plant and wait for a rain. Wait for rain and then plant if it's not too late.

Harvest

Have harvest equipment adjusted and fields ready in advance. When harvesting wheat, run a minimum of straw through the combine. This will prolong the life of the machine and usually result in cleaner grain. Check behind the combine for harvest losses (Table 3).

Table 3. Harvest loss on soybeans due to combine cutter bar height

Cutter bar height	Yield loss
2 inches	0.3 percent
4 inches	2.6 percent
6 inches	9.5 percent

Adjust the machine if necessary. Lost grain is lost yield.

Dryland Cropping Systems

FAO uses the term 'drylands' to describe climates with fewer than 120 days growing season. These are divided into 'arid drylands' with less than 75 days growing season, and 'semi-arid' areas which have from 75 to 119 days growing season (Table 4).

Table 4. Drylands in developing countries by region (million ha)

	Growth days			
	<74	75-119	Total drylands	Total area
Africa	488.0 (17%)	231.8 (8%)	719.8 (25%)	2878.1
Southwest Asia	72.6 (11%)	61.8 (9%)	134.4 (20%)	677.4
Southeast Asia	47.4 (5%)	54.9 (6%)	102.3 (11%)	897.6
Central America	62.3 (23%)	14.5 (5%)	76.8 (28%)	271.6
South America	114.6 (6%)	142.8 (8%)	257.4 (14%)	1770.2
East Asia	27.7 (3%)	70.4 (7%)	98.1 (10%)	954.6
Total	812.6 (11%)	586.2 (7%)	1388.8 (18%)	7449.5

Though most semi-arid lands border the arid or desert climates of the world, they support a diversity of cropping (mostly with annual crops). Because of this diversity their climate should be quantified and any variations within them delineated. Such drylands have either one or two wet seasons. Bimodal rainfall regimes are characteristically found in latitudes between the wet tropics and the unimodal more temperate, semi-arid zones. Their distribution, however, is governed more by pressure systems and land masses. For example, bimodal patterns are rarely found in east Asia.

In the tropics, the wet season(s) coincide with a high solar angle except for a relatively small area in east Brazil. The tropical drylands are commonly called monsoon (in Asia) or savannah (in Africa) although there is confusion over these terms. In temperate areas, there is usually a single wet season that coincides with a low solar angle (winter). These wet-and-dry climates are generally called Mediterranean and, on the drier margins, semi-arid.

This delineation of climates is based on calculations of the relative constraints of soil water, temperature and solar radiation on growth of temperate and tropical crops. Each of the indices ranges from zero (no growth) to 1 (climate optimal for growth). The water, temperature and radiation indices are multiplied together to calculate a growth index, which is an estimate of the overall suitability of the location for crop growth in a particular week or month of the year.

Hutchinson's climate classification, unlike more traditional, geographically-oriented classifications, reflects seasonal fluctuations of environmental factors related quantitatively to crop growth. Tropical drylands that support cropping have mean annual water indices as low as 0.29, but have a growth index (the product of the water, temperature and radiation indices) during the highest quarter (three-month period) greater than 0.3 (commonly 0.3-0.8). This is sufficient to support good growth of a single crop.

Zone I coincides with parts of the Aw, Bsh and Cwa types of Koppen and Trewartha and corresponds with the 1.3, 1.4, 1.5, 1.7 and 1.9 zones of Papadakis. Hutchinson's E zone, which is not as hot (Table 5) coincides with parts of Koppen's Cfa, Cwa and Bsh types and Papadakis' zones 1.7, 2, 4.2 and 5 in the tropics.

Table 5. Classification of climates: critical characteristics which Hutchinson et al. use to define 34 agro-climatic zones. TI19 and TI28 are the thermal indices for temperate crops (having an optimum temperature for growth of 19°C) and tropical crops (28°C), MI is water index and GI is the growth index derived from multiplying the monthly water, temperature and radiation indices

Group	Annual mean TI_{19}	Highest quarter GI_{19}	Annual mean TI_{28}	Highest quarter GI_{28}	Annual mean MI
Very cold					
A	0.09	0.16	0.00	0.00	0.82
Cold					
B1	0.27	0.57	0.04	0.10	0.90
B2	0.42	0.72	0.05	0.13	0.94
Cool, dry					
C1	0.49	0.13	0.25	0.06	0.23
C2	0.39	0.28	0.14	0.11	0.45
C3	0.56	0.31	0.27	0.10	0.45
C4	0.49	0.41	0.27	0.32	0.32
Cool, wet					
D1	0.38	0.64	0.17	0.40	0.59
D2	0.49	0.57	0.38	0.61	0.53
03	0.43	0.78	0.13	0.33	0.91
04	0.50	0.79	0.26	0.56	0.88
05	0.63	0.66	0.17	0.18	0.76
Warm, seasonally wet/dry					
E1	0.81	0.65	0.35	0.13	0.55
E2	0.79	0.37	0.37	0.09	0.37
E3	0.75	0.46	0.36	0.30	0.45
E4	0.79	0.30	0.56	0.34	0.32
E5	0.97	0.85	0.51	0.51	0.57

E6	0.80	0.19	0.47	0.12	0.17
E7	0.89	0.69	0.54	0.62	0.62
E8	0.90	0.51	0.60	0.47	0.29
Warm, wet					
F1	0.57	0.78	0.43	0.81	0.91
F2	0.74	0.77	0.66	0.86	0.89
F3	0.84	0.81	0.35	0.51	0.78
F4	0.94	0.85	0.50	0.66	0.87
F5	0.98	0.89	0.63	0.66	0.96
Warm to hot, very dry					
G	0.63	0.10	0.64	0.12	0.10
Hot, dry					
H	0.56	0.19	0.89	0.44	0.20
Hot, seasonally wet/dry					
11	0.51	0.48	0.96	0.87	0.52
12	0.53	0.40	0.81	0.82	0.38
13	0.62	0.60	0.78	0.87	0.64
14	0.90	0.80	0.82	0.78	0.65
Hot, wet					
J1	0.65	0.66	0.94	0.84	0.66
J2	0.59	0.62	0.96	0.88	0.82
J3	0.68	0.67	0.94	0.88	0.97

The tropical drylands have close affinities with temperate semi-arid climates where annual cropping is practised. In the latter areas, the growing season exceeds the FAO limit of 119 days, and is often as much as 150-190 days because the cropping period coincides with the season of low temperature and evaporation. The C climates of Hutchinson *et al.* are cool and dry while their cropping systems overlap with the dry margin of their D climates, which are cool but wetter.

Within these environments, the length of growing season and the kind of cropping system are determined not only by rainfall distribution but by soil water-holding characteristics and topographic position. The latter affects runoff to and runon from neighbouring land. The wetter regimes within the 119-day growing season allow some relay cropping on soils with favourable water-holding characteristics, but single cropping is more typical unless runon water is available. The time available for a cropping sequence at Hyderabad, India (average annual rainfall 760 mm), as affected by soil water-holding characteristics.

Because the drylands occupy a spectrum of water availability and cropping patterns, they are subdivided into two or more zones by Hutchinson *et al.* and others. For example, in the Bolivian lowlands (below 1000 m), Cochrane distinguishes four subclasses of rainfed cropping: very humid, humid, semi-humid and dry. Within these zones soil moisture is adequate for crop growth (at a 75% rainfall probability) for 8, 7, 5 and less than 5 months per year respectively. In West Africa, Kowal and Kassam distinguish seven bioclimatic regions within the savanna.

Semi-arid environments do not include those dry climates in which rainfall infiltration exceeds evapotranspiration and direct evaporation for so short a period that even the most rapidly maturing crops cannot be grown without irrigation. Hutchinson *et al.* consider their hot dry climate (H) too dry or very marginal for rainfed crops and their very dry climate (G) unquestionably cannot support cropping. This latter category supports dry savannah (tallgrass) vegetation in the tropics, dry temperate shortgrass steppe in temperate areas, and also includes hot deserts. Dryland cropping requires a rainy season of about 55 days, which corresponds to an average annual rainfall of less than 250 mm in temperate regions and 300-350 mm in the tropics. This, depending on soil water holding capacity, will support crop growth for 65-75 days.

The Wheat-Fallow (WF) system, practiced for many years in the semiarid Western Great Plains, was a definite improvement over continuous cropping. It stabilized yields and provided farmers with a reliable income from year to year. However, wheat fallow is an inefficient user of annual precipitation in a region where water is the major limiting factor. Data collected in the Great Plains show that typically less than 25 percent of the precipitation received during the 14-month fallow period is stored in the soil. When little or no residue is left on the soil surface, water storage efficiency is less than 20 percent.

Colorado and Nebraska research indicates that precipitation storage efficiencies of 40 to 60 percent is achieved when tillage is minimized or eliminated. The key to these improvements is maintaining crop residue on the soil surface and minimizing or eliminating soil disturbance. Unfortunately, little added wheat yield results from the additional water stored under no-till compared to mulch tillage systems. In fact, it costs more to save the additional water in a no-till system than the value of the added grain yield. Weed control with herbicides is more costly than with mechanical tillage.

Water savings in no-till systems can be converted to profit only by switching to more intense cropping systems where fallow time is decreased and summer Crops, like corn, grain sorghum, proso millet and annual forages are added to the rotation. Maximizing water storage, which permits more intensive rotations, requires residue maintenance on the soil surface to trap snow, absorb raindrop impact, slow runoff and minimize evaporation, as well as complete weed control at critical times in the production system.

Producers can use the following recommendations to improve precipitation use efficiency in eastern Colorado dryland farming systems, and improve soil, water and air quality while increasing net income. These steps are built around rotations where winter wheat is the starting point in the system.

Intensification of Dryland Crop Rotations

Step 1: Weed control

Weed control in growing wheat is the goal of this step. If controlled in the wheat, the stubble at harvest is essentially weed free. If weeds are present at harvest, apply the appropriate herbicide for that species . Apply residual herbicides in late July to mid-August to control volunteer wheat, downy brome and jointed goatgrass. Various combinations of herbicides with small amounts of atrazine are effective for this purpose.

Keep the atrazine rates low and within labelled amounts for your soils and climate. Failure results in long-term damage to wheat yields. Do not till in this phase of the rotation. Dollars spent for herbicides in this phase give maximum returns compared to any other point in the system. It should be possible to store 3 to 5 inches of water between wheat harvest to fall freeze-up if the stubble is kept weed free. By not tilling, the maximum amount of standing stubble can trap snow in the winter months and increase water storage by spring.

Step 2: Spring crop choices and cultural practices

Possible summer Crops to follow wheat in the rotation are corn, sorghum, proso millet, sudex or sunflowers. Plant corn in areas north of Cheyenne Wells and grain sorghum in southern areas. Proso millet is a good option in the northern area, but not in the southern areas where yields and markets are poor. Annual forages, like sudex and hay millet, are excellent choices for livestock producers. Whatever the summer crop, plant it into the wheat stubble or other residue base with minimal disturbance. A minimum till or slot planter is recommended. Apply an appropriate herbicide according to the label and use for weed control during the summer period. Avoid cultivation of row Crops unless the herbicide fails. Choose crop varieties that adapt to your elevation, latitude and longitude.

Step 3: Rotation choice

Specific rotation choices include: wheat-corn-fallow; wheat-sorghum-fallow; wheat-proso millet-fallow; wheat-corn-proso millet-fallow; etc. More intensive rotations (wheat-corn-proso millet-forage) are possible where no fallow is used. Consider any combination of Crops that adapt to your environment and fit your equipment, budget, and overall livestock feed requirements. Do not follow a particular crop with itself. There are too many insect, disease and volunteer problems when you repeat a species in a rotation.

Research suggests that corn follow wheat and proso millet, or an annual forage follow corn or grain sorghum. If you intend to keep winter wheat in your rotation, some fallow prior to planting winter wheat probably will be necessary. Another way to get back to winter wheat is to plant a short-season annual forage in the spring and harvest the forage prior to August 1.

If regrowth occurs, kill immediately with a contact (no residual) herbicide so that all precipitation received after August 1 is stored for wheat planting. Wheat should be planted no-till into the forage stubble. If it was a dry summer and no stored soil moisture is present, you may have to wait until spring and go back to another summer crop.

Step 4: Fertilization

Nitrogen, phosphorus and zinc are the nutrients of concern when making fertilizer decisions. Zinc deficiency on wheat has never been observed in Colorado, but it is common for corn and sorghum. Wheat is the crop most likely to respond to phosphorus fertilization. Base the amount of fertilizer to apply, if any, on soil test results. Ideally, collect separate samples from various soils on your farm. Sample eroded hillsides separately from summits and separately from lower-lying areas. Soil colour is a good way to judge whether soils should be sampled separately. Apply fertilizer according to each soil test if possible. This will result in the least-cost fertilizer programme.

Step 5: Overall considerations

All of the suggestions for intensifying cropping under dryland conditions are contingent upon maximum water capture and minimum losses to weeds, runoff and evaporation. No-till practices are required in certain phases of each system to be reasonably sure of conserving enough water to be successful. No-till is essential, for example, when going from wheat to corn or from corn to proso millet. It is feasible and most profitable, however, to substitute some tillage for herbicides in a summer fallow phase preceding wheat planting. An exception is a short fallow period from harvest ofa summer forage crop in August to a September wheat planting. No-till is the only feasible means of weed control in that situation.

Climate, Tillage, Weeds and Water Conservation

Climate

Most of the wheat producing areas are in the 15- to 18-inch precipitation zones of eastern Colorado. In these areas, rainfall may be extremely variable in total amount received per year and consistency throughout the growing season. Severe spring and summer storms may develop, releasing 2 to 3 inches of rain per hour. On the average, precipitation is higher in May, June, July and August (2 to 3 inches per month) and low (less than 1 inch per month) during the remaining season.

Tillage effects on water conservation

Assuming soil moisture is present, expect soil water losses from different tillage operations (Table 6). Implements that stir the soil, such as the one-way disk and chisels, cause more soil moisture losses, while sweeps and rod weeders cause less mixing and moisture loss.

Table 6. Water losses from different operations 1 and 4 days after tillage.

Operation	1 day	4 days
	--- inches of water ---	
One way	0.33	0.51
Chisel	0.29	0.48
Sweep plow	0.09	0.14
Rod weeder	0.04	0.22

Table 7. Percent residue lost after using different equipment.

Operation	% residue lost due to each operation
Spraying	0
Sweeps (24 inches)	10
Disk drills	20
Disk chisels	10
Rod weeder	15
Chisel plow - Straight points-Twisted Points	2550
Tandem disk 3" deep	80
Tandem disk 6" deep	90
Moldboard plow	90-100
Overwinter weathering	15-25

Eliminating runoff can significantly improve water infiltration. Applying water under various field conditions shows that no-till and high levels of residue on the soil surface reduce runoff compared to standard dryland tillage systems (Table 8). During this experiment, simulated rain was applied equally on all plots. Treatments were no-till, disked and disked with 3,000 pounds wheat straw per acre. Runoff began first on the no-till; however, runoff remained low throughout the experiment, compared to the disked treatment and the treatment with straw cover. The no-till and disked/straw cover allowed more water infiltration than the disked treatment.

Table 8. Water runoff when applied to no-till, disked and disked soil with straw cover.

Time (mins.)	No-till	Disked	Disked/straw
	--- Runoff inches/hour ---		
15	Started		
18	---	Started	
20	0.8	1.4	
25	2.1	3.0	Started
30	2.4	3.5	0.4
40	2.4	3.5	2.4

Soils with stubble cover here reduce wind velocities at the surface and temperatures, reducing evaporation from the soil surface. Experiments at Akron, Colorado indicate that water losses were 1.5 times greater on bare soil compared to soils with 3,000 pounds of wheat straw. As wind speed and temperatures increase and humidity decreases, water evaporation increases. With a 5 mph wind at Akron, Colorado, when temperatures are 40 degrees F, a water evaporation rate of 0.1 inches per day occurs. However, if temperatures reach 80 degrees F, water evaporation exceeds 0.3 inches per day. Any methods that reduce soil temperatures and wind velocity at the soil surface reduce water losses from the soil.

Plant Water Use

Under normal conditions, plants use water from the soil profile at rates depending on the species of the plant. Various Crops transpire water at rates between 300 and 900 pounds water per unit of dry matter (T:R) produced (Table 9). Some weeds are very inefficient and use much more water per unit of dry matter produced than agronomic Crops.

Table 9. Summary of the transpiration ratio of various Crops and weed species at Akron, CO.

Crop	T:R[1]	Crop	T:R[1]
Proso millet	267	Sunflower	630
Pigweed	287	Oats	583
Sorghum	304	Gumweed	608
Corn	349	Soybeans	646
Sugarbeets	377	Dry beans	700
Wheat	528	Rape	714
Cotton	568	Lambsquarters	801
		Ragweed	948

Pounds of water transpired per pound of above-ground dry matter produced. Water weighs 8.34 pounds/gallon.

Pounds of water transpired per pound of above-ground dry matter produced. Water weighs 8.34 pounds/gallon.

Water Use by Weeds

Weeds, just like cultivated Crops, extract water, usually in greater amounts per unit of dry matter than field Crops. They are competitive for water and nutrients. Problem weeds may differ from one farm to another. Constant tillage by disk implements decreases grassy weed competition, while blade implements increases grassy weed competition.

The most prevalent weeds in dryland fields of eastern Colorado include kochia, Russian thistle, pigweed, rye, downy brome and jointed goatgrass. Recently, goatgrass has invaded fields, causing nutrient and water losses, and discounts at the market place because of excessive foreign material. Goatgrass is a noxious weed.

Control of winter annual grasses in winter wheat-fallow rotations is extremely difficult because they germinate in the fall during wheat germination. Most of the seed matures and falls to the ground before wheat harvest. Selective weed control by chemicals has not been very successful. Winter annual grasses compete well in no-till systems and can result in failure if they get out of control. Intense rotations result in increased control of winter annual grasses because summer Crops are included in the system.

6

Soil - Water Conservation

Soil conservation is a set of management strategies for prevention of soil being eroded from the earth's surface or becoming chemically altered by overuse, salinisation, acidification, or other chemical soil contamination. The principal approaches these strategies take are:

— choice of vegetative cover

— erosion prevention

— salinity management

— acidity control

— encouraging health of beneficial soil organisms

— prevention and remediation of soil contamination

— mineralisation

Many scientific disciplines are involved in these pursuits, including agronomy, hydrology, soil science, meteorology, microbiology, and environmental chemistry. Decisions regarding appropriate crop rotation, cover crops, and planted windbreaks are central to the ability of surface soils to retain their integrity, both with respect to erosive forces and chemical change from nutrient depletion. Crop rotation is simply the conventional alternation of crops on a given field, so that nutrient depletion is avoided from repetitive chemical uptake/ deposition of single crop growth.

Cover crops serve the function of protecting the soil from erosion, weed establishment or excess evapotranspiration; however, they may also serve vital soil chemistry functions. For example, legumes can be ploughed under to augment soil nitrates, and other plants have the ability to metabolise soil contaminants or alter adverse pH. The cover crop Mucuna pruriens (velvet bean) has been used in Nigeria to increase phosphorus availability after application of rock phosphate. Some of these same precepts are applicable to urban landscaping, especially with respect to ground-cover selection for erosion control and weed suppression.

Evergreen species are preferred to achieve year-round protection; however, as long as foliage is present in the seasons of bare soil surfaces, the effect of deciduous trees may also be adequate. Trees, shrubs and groundcovers are also effective perimeter treatment for soil erosion prevention, by insuring any surface flows are impeded. A special form of this perimeter or inter-row treatment is the use of a "grassway" that both channels and dissipates runoff through surface friction, impeding surface runoff, and encouraging infiltration of the slowed surface water.

The problem of conserving soil and moisture is also of very great importance in the extensive regions of low and uncertain rainfall, forming parts of Punjab, Madhya Pradesh, Maharashtra, Andhra Pradesh and Karnataka. These tracts are characterised by scanty, ill-distributed and highly erosive rains, undulating topography, high wind velocity and generally shallow soils. The period of heavy downpours from August to October is the period of the heaviest erosion in these regions.

Wind erosion also has been responsible for destroying the valuable top soil in many areas. Halting the march of desert in Rajasthan is one of the vital and outstanding problems facing the country today. An extreme example of sand movement from the coast is to be seen in the Saurashtra region of Gujarat where the once-flourishing ports are now covered with advancing sand-dunes.

In addition to the erosion of the cultivated fields, neglected pastures and wastelands, considerable roadside erosion also takes place owing to the defective highway engineering. Defective drainage and water-logging throw appreciable areas out of cultivation every year and indirectly increase the erosion hazards.

Extent of Soil Erosion

In India, there is very little area free from the hazard of soil erosion. It is estimated that out of 305.9 million hectares of reported area, 145 million hectares is in need of conservation measures. Severe erosion occurs in the sub-humid and per-humid areas due to high rainfall and improper management of land and water. Agricultural land in the major part of the country suffers from erosion. Apart from reducing the yields through the loss of nutrients, erosion destroys the soil resources itself every year.

For example, in Maharashtra over 70 per cent of the cultivated land has been affected by erosion in varying degrees and 32 per cent of the land having been highly eroded is no longer cultivable. In the Sholapur district, nearly 17 per cent of the land of medium depth (more than 45 cm) has deteriorated into shallow soils (less than 45 cm) in 75 years from 1870 to 1945. Similarly, in Akola, Buldana and Yeotmal districts, the number of fields with less than 37.5 cm soil depth increased during the same period by 54, 16 and 8 per cent respectively. As much as 2.3 million ha is already under ravines scattered all over India. The ravines apart from ruining the soil resources for ever are a constant threat to the adjoining fertile cultivated lands.

Table 1 . Problem of conservation of land and water in India

Particulars	Area in million hectares	
	Total Area	Soil-Conservation problem area
Forest	61.170	20
Culturable wasteland	17.362	15
Permanent pastures and other grazing land	14.809	14
Land under miscellaneous tree crops and groves	4.218	1
Fallow lands :		
(i) Fallow lands other than current fallows	9.168	8
(ii) Current fallows	11.132	7
	20.5	15
Net area under cultivation	137.9	80
Other land uses, not available for agriculture, forest,etc.	50.188	--
	305.947*	145

*305.947 million hectares is the reported area for land-utilisation statistics out of a geographical area of 328.809 million hectares

The denudation of forests and vegetation in the Shivalik Hills, the Himalayas, the Western Ghats, the Eastern Ghats and other mountain ranges of the Deccan have resulted in floods and chos (rainy season torrents) which destroy good agricultural land. For instance, the chos in the Hoshiarpur district of Punjab covered an area of 19,282 ha which increased to 32,022 ha in 1884, to 37,730 ha in 1897 and to about 60,000 ha in 1936.

In the Himalayan regions, landslides and landslips are very serious problems caused by improper land management. The recent landslide disaster in the winter of 1969 in the Darjeeling district of West Bengal is a reminder of the bad management of land resources and a portent of worse things to come. The erosion problem along the mountain roads is assuming very serious proportions. The Border Road Organisation is finding this problem so acute that a National Seminar was organised on the problems of controlling erosion and stabilising slopes along the highways. A huge amount of money is spent every year for keeping these important lines of communication open.

Costly reservoirs constructed under the river-valley projects are being silted up at an alarming rate owing to the denudation of forest vegetation, the cultivation of steep slopes without adopting any conservation practices, landslides, torrents, etc. Morever, as the pressure on land increases there will be a tendency and a demand for opening up marginal and steep lands for cultivation. These lands will be in greater need for measure to conserve soil and water.

In fact, if erosion is permitted to continue at its present rate, it is possible that all work will be the reclamation of soil rather than the conservation and management of soil and water. Soil erosion is the wearing away of land surface by the action of such natural agencies as water and wind.

The following types of erosion are observed in India :

Normal or geologic erosion: This is a normal feature of any landscape. Geologic erosion takes place steadily but so slowly that ages are required for it to make any marked alteration in the major features of the earth's surface. There is always an equilibrium between the removal and formation of soil, so that unless the equilibrium is disturbed by some outside agency, the mature soil preserves more or less, a constant depth and character indefinitely.

Accelerated soil erosion: The removal of the surface soil from areas denuded of their natural protective cover as a result of human and animal interference takes place at a much faster rate than that at which it is built up by the soil-forming process. This accelerated detachment rapidly ravages the land and it is with this type of soil erosion that we are so seriously concerned. Nature requires, on an average, about 1000 years to build up 2.5 cm of top soil, but wrong farming methods may take only a few years to erode it from lands of average slope.

Wind erosion: Wind erosion takes place normally in arid and semi-arid areas devoid of vegetation, where the wind velocity is high. The soil particles on the land surface are lifted and blown off as dust storms. When the velocity of the dust-bearing winds is retarded, coarser soil particles are deposited in the form of dunes and thus fertile lands are rendered unfit for cultivation. In other places, fertile soil is blown away by winds and the subsoil is exposed, as a result the productive capacity of the soil is considerably reduced.

Water erosion: Soil erosion caused by water can be distinguished in three forms, viz. (1) sheet erosion, (2) rill erosion, and gully erosion.

Sheet erosion: Sheet erosion removes a thin covering of soil from large areas, often from entire fields, more or less, uniformly during every rain which produces a run-off. This type of erosion is very insidious, since it keeps the cultivator almost ignorant of its ill-effects. It is generally neglected, although the soil deteriorates slowly and imperceptibly. Its existence, however, can be detected by the muddy colour of the run-off from the fields.

Gully erosion: When rill erosion is neglected, the tiny grooves develop into wider and deeper channels, which may assume a huge size. This is called 'gully' erosion. Gullies are the most spectacular evidence of the destruction of soil. The gullies tend to deepen and widen with every heavy rainfall. They cut up large fields into small fragments and, in course of time, make them unfit for cultivation.

Landslides or slip erosion: A landslide is defined as an outward and downward movement of the slope-forming material, composed of natural rocks, soil, artificial fills, etc. The fundamental causes of landslides are topography of the region and geological structure, the kinds of rocks and their physical characteristics. The immediate cause of a slide may be an earthquake, or a heavy rainfall, which unduly saturates the ground or a part of a road. However, these are accidents rather than fundamental causes.

Stream-bank erosion: Torrents are defined as hill streams characterised by wide-spreading beds on emergence from the hills with ill-defined banks, flashy flows and swift currents. Usually, they are dry water-courses, except during the rainy season when with every downpour in their catchment, they get very much swollen with flood and subside almost to its normal tiny size immediately after the storm is over.

These sudden and violent flows are responsible for moving immense quantities of detritus, comprising boulders, shingles, sand and silt, depending upon the geology of the terrain. This debris gets deposited in the torrent bed in the form of scattered islands owing to the sudden widening of the torrent channel after it emerges from the hills, or owing to the flattening of the gradient in the lower reaches, or because of obstructions caused by wild vegetation and uprooted trees. The bed level of the torrent is raised by these deposits. These deposits, in turn, reduce the transporting capacity of the torrent, resulting in overflows and the meandering of the course and in the erosion of the banks.

Mechanism of Erosion

Water Erosion

Soil erosion caused by rainfall is the result of the application of energy from two distinct sources, namely (i) the falling rain drops, and (ii) the surface flow. The energy of a falling rain drop is applied slantingly or vertically from above, whereas that of a surface flow is applied more or less horizontally along the surface of the ground. The chief role of the falling rain drop is to detach soil particles, whereas that of the surface flow (outside the rills and gullies) is to transport the soil. The falling rain drop also makes a major contribution to the movement of the soil on unprotected sloping lands during the period of heavy-impact storms, by splashing large quantities downslope and by imparting transporting capacity to the surface water by keeping it turbid. More than 100 tonnes of soil per hectare can sometimes be lost yearly in this fashion from a bare and highly detachable soil on slopping land.

Wind Erosion

Wind is responsible for three types of soil movement in the process of wind erosion. They are known as : (i) saltation, (ii) suspension, and (iii) surface creep.

Saltation: The major portion of the soil carried by the wind is moved in a series of short bounces called "saltations". The soil carried in a saltation consists of fine particles ranging from 0.1 to 0.5 mm in diameter. saltation is caused by the direct pressure of wind on soil particles and their collision with other particles. After being pushed along the ground surface by the wind, the particles leap almost vertically in the first stages of saltation. Some grains rise only a short distance; others leap 30 cm or higher, depending on the velocity of the rise from the ground.

Suspension: Very fine soil particles, less than 0.1 mm in diameter, are carried into suspension, being kicked up into the air by the action of particles in saltation. The movement of fine dust in suspension is completely governed by the characteristic movement of the wind. Suspended material is carried long distances from its original location and is thus a complete loss to the eroded area, especially when erosive winds are from different directions.

Surface creep: Soil particles, larger than about 0.5 mm in diameter but smaller than 0.1 mm, are too heavy to be moved in saltation but are pushed or spread along the surface by the impact of particles in saltation to form a surface creep.

About 90 per cent of the total soil movement in wind erosion is below the height of 30 cm, and about 50 per cent of it is within 5 cm of the ground level. The control of wind erosion is mainly based on the reduction or elimination of movement in saltation.

Influencing Factors of Erosion

The following factors are influencing erosion.

i) precipitation (its intensity and amount),

ii) the slope of the land (its degree and length),

iii) the type of soil, and

iv) the nature of the ground cover and land use.

Precipitation

Precipitation is the most important factor influencing soil erosion. The intensity of rainfall, its duration and frequency influence the rate and the volume of run-off. A light rain which can be easily absorbed in the soil causes no run-off and soil loss. As the intensity of rain increases and more rain falls than can enter the soil (infiltration), there is of run-off and soil loss. Rainfall of long duration and greater frequency increases both the total run-off and soil loss. Apart from the intensity and duration of rainfall, the soil moisture is also important in determining the run-off and soil loss by erosion. If the soil is already saturated with water, the same amount and intensity of rainfall will cause more run-off and soil loss from it than from a dry soil.

Degree and length of slope

The speed and the extent of run-off depend on the slope of the land. The greater the slope, the greater is the velocity of the flow of the run-off. According to the law of falling bodies, velocity varies as the square-root of the vertical drop. Hence, if the land slope is increased four times, the velocity of the water flowing on the slope is approximately doubled. If the velocity of the run-off water is doubled, its energy, i.e. erosive power, is increased four times, as the latter varies as the square of the velocity. Similarly, the quantity of the material of a given size that can be carried is increased about thirty-two times (varies as the fifth power of the velocity), and the size of the particles that can be transported by pushing or rolling is increased about sixty-four times (varies as the sixth power of the velocity).

There is thus a rapidly increasing rate of soil loss as the slope of the field becomes steeper. The erosion hazard is not simply added but is multiplied as the field extends back on the steeper part of the farm.

Type of soil

The type of soil, i.e. structure, texture, organic matter content, its infiltration capacity and permeability, greatly affects the soil loss and run-off. Fine soils are more susceptible to erosion than coarse soils, since rain-water enters in and passes through a dense clay much more slowly than through a porous sand or gravelly soil. In India, it has been observed that deep lateritic soils at Ootacamund and red soils at Deochanda have the lowest rate of run-off; the alluvial soils at Vasad and Dehra Dun have a very high rate of run-off; the black soils have an intermediate rate of run-off, but still the rate of run-off is high. Lateritic clays are less erodiable. The soil left in loose and pulverised condition is particularly liable to erosion through sheet-wash and gullying.

Nature of ground cover and land use

When rain falls on a surface covered by a thick mantle of plants, its velocity and erosive power are reduced and most of the water either quickly percolates through the soil or moves over the surface with non-erosive velocity. Areas not protected with thick cover of plants are unable to absorb water effectively, because the dashing rains shatter the soil surface, the fine soil particles go into suspension and the thick mixture of water and soil quickly fills and closes the tiny interstices in the soil, reducing infiltration and consequently increasing run-off and soil loss.

The above data on run-off and soil loss under different soil, climatic and slope conditions clearly indicate that if the land is left undisturbed under a natural cover, the run-off and soil loss are the least; the soil loss and run-off increase steeply when the vegetation is removed and the land is cultivated.

Effect of land use on run-off and soil erosion

Treatment	Rainfall causing run-off (mm)	Run-off (mm)	Soil loss (tonnes/ha)
Alluvial soils - 8% slope - Dehra Dun			
Bare Fallow	1,223	339	42.4
Cultivated Fallow	1,223	889	156.0
Natural grasses	1,223	265	1.0
Maize-wheat (up-and-down cultivation)	1,223	670	28.5
Black soil - 0.5% slope - Kotla			
Natural cover	657	33	0.3
Cultivated fallow	657	111	3.5
Jowar (kharif)	657	79	2.9
Black soil - Sholapur			
Natural vegetation	342	28	1.3
Vegetation removed	342	118	44.3
Johar (rabi)	342	112	86.5
Red soil - 2% slope - Deochanda (D.V.C.)			
Natural fallow	1,002	105	0.6
Overgrased fallow	1,002	222	3.3
Maize (contour cultivation)	1,002	64	3.3

Plant-nutrient losses due to soil erosion - Kanpur

Nutrient (kg/ha) lost in run-off of water and eroded soil

Degree of slope	Organic matter	Total N	P_2O_5	K_2O	CaO	MgO
0.5%	86.8	5.8	10.7	42.8	53.4	41.4
1.5%	92.8	6.5	11.1	52.9	59.2	78.5
3.0%	173.9	10.8	23.5	117.8	203.2	211.8

Alongwith the loss of run-off of water and soil, considerable amount of plant nutrients are also lost. The loss of plant nutrients in alluvial soil at Kanpur increased with the increase in the degree of the slope and the increase was very steep when the degree of slope increased from 1.5 per cent to 3.0 per cent.

Measures of Soil and Water-Conservation

The key to soil and water conservation is the utilisation and treatment of land according to its capability.

Land-capability classification. Any soil and water-conservation project includes two distinct sets of operations, viz. (1) the mapping of land for classification according to its capability, and (2) planning and executing measures to check erosion, improve land productivity and reclaim wasteland. The farm plans for effective soil and water conservation are based largely on the capability of the land. The land-capability classification map is normally prepared by interpreting a standard soil-survey map.

Land-capability classification is a systematic arrangement of different kinds of lands according to those properties that determine the ability of the land to produce crops on a virtually permanent basis. These are the major soil characteristics of the land, e.g. the texture of the top soil, its effective depth, permeability of the top soil and subsoil, and associated land features, e.g. the slope of the the land, the extent of erosion, the degree of wetness and susceptibility to overflowing and flooding. The grouping of soils into capability classes is done primarily on the basis of their capability to produce common cultivated crops and pasture plants without deterioration over a long period.

The land-capability classes are based on the intensity of hazards and the limitations of use. The land-capability classes range from the best and most easily farmed land to that which has no value for cultivation, grazing or forestry, but which may be suited to wild-life, recreation or for watershed protection. They all fall into 2 broad groups: one suitable for cultivation and other land uses, and the other not suitable for cultivation, but suitable for other land uses.

Land for Cultivation

Class I (Green colour). Soils in class I have very few or no limitations that restrict their use. This type of land is nearly level and the erosion hazard is low. The soils are deep, well-drained, easily worked, hold water well and are either fairly well supplied with plant nutrients or highly responsive to the application of fertilisers. The soils are not subject to damage because of overflow. The local climate must be favorable for growing many of the common field crops. In irrigated areas, the soils may be in class I, if the limitation of the arid climate has been removed by relatively permanent irrigation works.

These soils need ordinary management practices to maintain productivity. Such practices may include the use of one or more of the following: fertilisers, lime, cover and green-manure crops, conservation of crop residues and crop rotations. Soils in this class are suited to a wide range of plants, may be used for cultivated crops, pastures, forests, and wildlife, food and cover.

Class II (Yellow Colour). Soils in class II have some limitations which reduce the choice of plants or require simple conservation practices. The limitations of soils in class II may result from the effects of one or more of the following factors: (i) a gentle slope, (ii) a slight susceptibility to erosion, (iii) less than ideal soil depth, (iv) occasional damaging overflow, (v) wetness which can be corrected by drainage, but existing permanently as a moderate limitation, (vi) slight to moderate salinity or sodium, easily corrected but likely to recur, and (vii) a slight climatic limitation on soil use and management.

These soils require careful management. The limitations are only a few and the practices are easy to apply. They may need one or more of the following practices: terracing, strip cropping, contour cultivation, water disposal area, covered with vegetation crop rotation, cover and green-manure crops, stubble mulching, the use of fertilisers, manure and lime. These soils may be used for growing cultivated crops, raising pastures, forests, and for wild-life, food and cover.

Class III (Red Colour). Soils in class III have moderate limitations which reduce the choice of plants or require special conservation practices. Soils in class III have more restrictions than those in class II and, when used for cultivated crops, the conservation practices are usually more difficult to apply and to maintain.

Limitations of soils in class III may result from the effects of one or more of the following factors: (i) a moderately sloping land, (ii) moderately susceptibility to water or wind erosion, (iii) frequent overflow accompanied with some crop damage, (iv) very slow permeability of the sub-soil, (v) wetness or continuing water-logging after drainage, (vi) shallow soil depth up to the bed-rock, hard-pan or clay-pan which limits the rooting-zone and water storage, (vii) low moisture-holding capacity, (viii) moderate salinity or sodium, and (ix) moderate climatic limitations.

Class IV (Blue Colour). Soils in class IV have severe limitations that restrict the choice of plants and require careful management. The restrictions in the use of these soils are greater than those in class III and the choice of plants is more limited. When these soils are cultivated, very careful management is required and the conservation practices are more difficult to apply and to maintain.

The use of these soils for cultivated crops is limited as a result of the effect of one or more of the permanent features, such as (i) steep slopes, (ii) severe susceptibility to water or wind erosion, (iii) severe effect of past erosion, (iv) shallow soil, (v) low moisture-holding capacity, (vi) frequent overflow accompanied with severe crop damage, (vii) excessive wetness or continuing hazard of water-logging after drainage, (viii) severe salinity or sodium, and (ix) moderately adverse climate. These soils can be used for crops, pastures, forests, and wild-life food and cover.

Class V (Dark Green or Uncoloured). Soils in class V have little or no erosion hazard, but have other limitations, the removal of which is not practicable. They are used largely for pastures, forests, and wild-life food and cover. Such land is nearly level and is not subject to more than slight wind or water erosion. Cultivation is not feasible because of one or more limitations, such as overflow, stoniness, wetness or severe climate.

Examples of class V land are (i) soils of lowlands subject to frequent overflows which prevent the normal production of cultivated crops, (ii) nearly level soils with growing season that prevents the normal production of cultivated crops,(iii) the level or nearly level stony or rocky soils, and (iv) ponded areas where drainage for cultivated crops is not feasible but where soils are suitable for grasses or trees. Soils in class V are not suitable for raising cultivated crops, but are suitable for perennial vegetation (grazing and forestry, with few or no limitation). Pastures can be improved, and benefits from proper management can be expected. Physical conditions of soils are such that it is practicable to apply pasture improvements, if needed, such as seeding, liming, fertilising, and water control with contour furrows, drainage ditches, diversions of water spreaders.

Class VI (Orange Colour). Soils in class VI have severe limitations that make them unsuitable for cultivation and limit their use largely for pastures, or forests, or wild-life food and cover. Soils in class VI have continuing limitations which cannot be corrected, such as (i) steep slope, (ii) very severe erosion hazard, (iii) very severe effect of past erosion (iv) stoniness, (v) shallow rooting-zone, (vi) excessive wetness or overflow, (vii) low moisture capacity, (viii) salinity or sodium, and (ix) severe climate. Soils in this class are subject to moderate limitations under grazing or forestry use.

Class VII (Brown Colour). Soils in class VII have very severe limitations that make them unsuitable for cultivation and restrict their use largely to grazing, or forestation, or wild-life food and cover. The soils in this class are subject to severe limitations or hazards under either grazing or forestry use. The physical condition of soils is such that it is not practicable to adopt pasture improvements and water-control practices. Soil restrictions are sever e than those in the case of class VI soils.

Class VIII (Purple Colour). Soils and land forms in class VIII have limitations that preclude their use for commercial plant production and restrict their use to recreation, wild-life food and cover or to water-supply, water shed protection or for aesthetic purposes. Significant return on site benefits from soils and land forms in class VIII cannot be expected from management of crops, grasses or trees, although indirect benefits from wild-life, watershed protection or recreation may be possible.

Limitations which cannot be corrected may result from the effects of one or more of the following factors: (i) erosion or erosion hazard, (ii) severe climate, (iii) wet soil, (iv) stones, (v) low moisture capacity, (vi) salinity or sodium. Bad lands, rock outcrops, sandy beaches, marshes, deserts, river wash, mine tailings and other nearly barren lands are included in class VIII in order to protect other more valuable soils to control water or for wild-life or for aesthetic reasons.

The land-capability class is indicated on the maps by roman numerals I to VIII or by standard colors or by both.

Land-capability Subclass

A land-capability class is determined by the degree of limitations in land use together with the hazards involved. For example, in class III land, we have land suitable for cultivation but subject to moderate hazard of water erosion because of a steep slope or a moderate hazard of wind erosion on smooth land or moderate hazard of water-logging or overflow and a shallow depth to the bed-rock. Each of these kinds of limitations are recognised at the subclass. Four kinds of limitations are recognised at the subclass level.

Subclass (E) Erosion. It is made up of soils where the susceptibility to erosion is the dominant hazard in their use. Susceptibility to erosion and damage due to past erosion are major factors that govern the placing of soils in this class.

Subclass (W) Excess water. It is made up of soils where excess water is the dominant hazard in their use. Poor soil drainage, wetness, high water-table and overflow are the criteria for determining which soils belong in this subclass.

Subclass (S) Soil limitations within the rooting zone. It includes soils which have such limitations as the shallowness of the rooting-zone, stones, low moisture-holding capacity, low fertility difficult to correct and salinity or alkalinity.

Subclass (C) Climatic limitations. It includes soils where the climate (temperature or lack of moisture) is the only major hazard of limitations in their use.

The land-capability subclass is designated by the small letters which follow the land-capability classes. Roman numerals, e.g. IIe, IIIs, IVw capability class I has no subclass. Where two kinds of limitation can be modified or corrected and are essentially equal, the subclasses have the following priority e, w, s, c. where soils have two kinds of limitations, the dominant one is shown first.

Soil-water Conservation on Agricultural Land

Broadly speaking, the practical methods of soil and water conservation fall into two important classes, viz. Agronomic measures and mechanicla measures.

Measures for Agronomic Soil-Water Conservation

Agronomic practices for soil and water conservation help to intercept rain drops and reduce the splash effect, help to obtain a better intake of water rate by the soil by improving the content of organic matter and soil structure, help to retard and reduce the overland run-off through the use of contour cultivation, mulches, dense-growing crops, strip-cropping and mixed cropping.

Contour-farming

During intense rain storms, the soil cannot absorb all the

rain as it falls. The excess water flows down the slope under the influence of gravity. If farming is done up and down the slope, the flow of water is accelerated, because each furrow serves as a rill. The major part of the rain is drained away without infiltrating into the soil. The top fertile soil, along with plant nutrients and seeds, is washed off. All this results in a scanty and uneven growth of a crop.

A simple practice of farming across the slope, keeping the same level, as far as possible (which is technically called contour-farming) has many beneficial effects. The ridges and the rows of the plants placed across the slope form a continual series of miniature barriers to the water moving over the soil surface. The barriers are small individually, but as they are large in number, their total effect is great in reducing run-off, soil erosion and loss of plant nutrients. It has been experimentally proved that contour-farming reduces run-off and prevents soil erosion as compared with the up-and-down cultivation in the major groups of soils in India, viz. Alluvial soils, black soils and deep lateritic soils.

Apart from conserving the water and soil, contour-farming conserves soil fertility and increases crop yields. Contour-farming on alluvial soil, 2.2 per cent slope at Kanpur has conserved 11.3 kg of N, 11.7 kg of P_2O_5, 44.4 kg of K_2O, 398.1 kg of CaO, 118.1.kg of MgO in one season alone. These nutrients converted into fertilisers amount to 56.5 kg of sulphate of ammonia, 70 kg of single super-phosphate and 74 kg of muriate of potash per hectare in one season.

Contour-farming has also given 490 kg of johar grain and 273 kg of johar stalks more than the up-and-down cultivation at Kanpur. Thus every mm of rain-water conserved by contour-farming gave 22.5 kg of johar grain and 12.5 kg of johar straw more than the up-and-down cultivation. It has been proved that much less power is required to be exerted by man, animals and machines, if the cultivation is done in the contour instead of up-and-down the slope. There is less wear and tear of the implements and the same job is done in less time when contour-farming is practiced.

On long slopes, bunding is usually done to reduce the length of the slope. These bunds will serve as a good guide for contour-farming. All the cultural operations have to be done parallel to these bunds. On gentle slopes (between 0.5 to 2.0 per cent) bunding may not be essential. Contour (the line passing through the points having the same level) guidelines can be marked with the help of a hand level. On uniform slopes, these lines are to be marked about 50 m apart. Farming is done parallel to these lines.

The establishing of contour-farming on undulating land (having many depressions and ridges) is somewhat tedious. The water from each furrow collects in the depressions and results in breaches. The depressions are required to be filled up by levelling or may be left under grass.

Mulching

Surface mulches are used to prevent soil from blowing and being washed away, to reduce evaporation, to increase infiltration, to keep down weeds, to improve soil structure and eventually to increase crop yields. Inter-culture kills weeds and produces a five or seven cm thick soil mulch which helps to reduce evaporation from the top soil. It also breaks the surface crust which forms after each downpour. Studies on mulching carried out in India under rain-fed agriculture have concentrated on the measurement of crop responses rather than on the manner in which crop responses are influenced.

At Bellary, mulching with different materials did not give encouraging results in increasing the yields of cotton and johar crops. Mulching with straw og Encap Esso mulch increased the infiltration rate, improved soil moisture and increased the yields of wheat, barley, gram and linseed, succeeding the maize crop on eroded soils at Ranchi.

Growing of crops which provide the maximum cover, reduce run-off and soil loss. Cultivated legumes, in general, furnish a better cover and hence better protection to cultivated land against erosion than ordinary cultivated crops. The crops and the cropping systems will naturally vary from region to region, depending on the soil and climatic conditions.

7

Organic Agriculture

After half a century of high intensive input agriculture, the yield gap between best practices and farmers' fields remains large, agricultural lands continue to shrink and global environmental threats are a reality, e.g. erosion of biodiversity, desertification, climate change and other transboundary pollution. Agriculture intensification contributes to over 20 percent of global anthropogenic greenhouse gas emissions.

Agricultural activities affect 70 percent of all threatened bird species and 49 percent of all plant species. Uniform cultures have dramatically reduced the number of plants and animals used in agriculture; currently, 1 350 breeds face extinction, with two breeds being lost each week. Biodiversity erosion is exacerbated by the loss of forest cover, coastal wetlands and other wild relatives, important for the development of biodiversity and essential for food provision, particularly in times of crisis.

Despite FAO's and other institutions' efforts for global food security, the benefits of increased agricultural production often bypass the poorest sections of the world population. Over 450 million farmers have never had access to Green Revolution technologies such as mechanisation, irrigation, improved seeds and breeds, and synthetic fertilisers and pesticides. The World Food Summit stressed that hunger is a problem of access to means of food production or means to purchase food.

Although agriculture remains the world's single largest employer, rural economies suffer from sharp decline in real food prices and loss of entrepreneurial capacity. The terms of international exchange favour importations at the expense of local production. Developing countries, which traditionally have had a net surplus in agricultural trade, increasingly depend on food imports. The conventional food production model ties farmers into conditions of dependence on large corporations to buy agricultural inputs (seeds, fertilisers, pesticides) and to sell their produce.

Need a Paradigm Shift

Protecting biodiversity at genes, species and ecosystem levels through germplasm banks and protected areas is not sufficient. The maintenance costs of gene banks are high, up to half of the material collected is in need of regeneration, and "freezing" genetic resources denies their evolution. Biodiversity is best maintained through sustainable utilisation and selection by food providers. Animals move across boundaries and ecosystems are not immune to air and water pollution. The 12 percent of global land areas "fenced" for nature protection are located within or around a 40 percent land surface used by agriculture and forestry. Food systems should be viewed as an integral part of the ecosystem. "There is a need to manage agricultural land as part of a larger landscape that explicitly considers ecological functioning".

Rural communities need land to derive livelihoods. National parks and protected areas have always had local culture as components of the landscape. Ecotourism management by local residents generates supplementary income for both management of protected areas and rural livelihoods. Agritourism creates new opportunities for farm employment, specialty foods and safe agro-ecosystems. There is a need to reconnect farmers to the land by linking their productive activities with nature conservation. A symbiotic relationship between agriculture and natural landscape exists in the presence of ecologically managed systems.

Poor farmers and market-marginalised communities need a food production model that relies on local natural resources and ecological management. Where food surpluses are generated, markets should guarantee "fair" prices. Managing local resources without having to rely on external inputs involves substituting purchased goods by knowledge of natural processes that optimise competition for nutrients and space between species within the agro-ecosystem. By stressing diversification and adaptive management, agricultural systems can improve soil and water quality and the ecological services that support agriculture, while significantly decreasing vulnerability to weather vagaries or other factors. Where markets for ecologically-produced food exist, farmers' return on labour are higher through consumers' willingness to pay higher prices.

Organic Agriculture and International Agreements

Organic agriculture offers a means to address food self-reliance, rural development and nature conservation. The common thread in this ambitious goal is the sustainable use of biodiversity; in terms of both agriculture contribution to biodiversity and biodiversity contribution to agriculture. To be successful, organic agriculture needs functional groups of species and essential ecosystem processes as its main "input" to compensate for the restriction on (or lack of) synthetic input use.

In fact, a close relationship exists between organic agriculture and the maintenance of biodiversity. This is expressed in the rules and regulations that govern certified organic agriculture and by the practical experiences of organic farmers around the world. Selected international codes and agreements expressing the organic agriculture-biodiversity linkages and need for further consideration are the following:

— The Vignola Declaration and Action Plan on the Relationship Between Nature Conservation, Biodiversity and Organic Agriculture of the International Federation for Organic Agriculture Movements, the World Conservation Union and the World Wide Fund for Nature. This Declaration recognises that "organic agriculture is essential for conserving biodiversity and Nature" and "asks environmentalists, farmers, politicians, industry and international institutions to support and develop organic agriculture as the most ecologically-sound agricultural system".

— The FAO/WHO Codex Alimentarius guidelines for Organic Food define organic agriculture as "a holistic production management system which promotes and enhances agro-ecosystem health, including biodiversity, biological cycles, and soil biological activity. The primary goal of organic agriculture is to optimise the health and productivity of interdependent communities of soil life, plants, animals and people".

— The Convention on Biological Diversity (CBD) "encourages the development of technologies and farming practices that not only increase productivity, but also arrest degradation as well as reclaim, rehabilitate, restore and enhance biological diversity and monitor adverse effects on sustainable agricultural biodiversity" such as "*inter alia*, organic farming". Furthermore, the CBD Global Strategy for Plant Conservation includes outcome-oriented global targets for 2010 whereby using plant diversity sustainably considers, among others, organic food as an indicator of direct measures to reach the target of "30 percent of plant-based products derived from sources that are sustainably managed".

— The Ninth Session of the Commission on Genetic Resources for Food and Agriculture recommended that FAO further studies the "contribution of genetic resources to the economies of member countries, the impact of agricultural subsidies on biodiversity, and the competitiveness of organic agriculture".

Organic Agriculture in Enhancing Biodiversity

In organic agriculture, biodiversity is both instrument and aim. Natural ecological balance, below and above ground, is key to its success. A healthy soil is the base for food production and a diversity of plants and animals on land prevents pest and disease outbreaks. Although organic agriculture is committed to the conservation and enhancement of biodiversity, many systems today remain limited to input substitution. To be unlocked, the real potential of organic agriculture on biodiversity requires a stronger shift to a systems approach, based on improved understanding of ecosystem functions.

Living Soils for Agriculture

Soils contain enormous numbers of diverse living

organisms assembled in complex and varied communities. Soil biodiversity reflects the variability among living organisms in the soil - ranging from the myriad of invisible microbes, bacteria and fungi to the more familiar macro-fauna such as earthworms and termites. Plant roots can also be considered as soil organisms in view of their symbiotic relationships and interactions with other soil components. These diverse organisms interact with one another and with the various plants and animals in the ecosystem, forming a complex web of biological activity. Environmental factors, such as temperature, moisture and acidity, as well as anthropogenic actions, in particular, agricultural and forestry management practices, affect to different extents soil biological communities and their functions.

Soil organisms contribute a wide range of essential services to the sustainable functioning of all ecosystems. They act as the primary driving agents of: nutrient cycling, regulating the dynamics of soil organic matter, soil carbon sequestration and greenhouse gas emissions; modifying soil physical structure and water regimes; enhancing the amount and efficiency of nutrient acquisition by the vegetation; and enhancing plant health. These services are not only critical to the functioning of natural ecosystems but constitute an important resource for sustainable agricultural systems.

Healthy Soils

Capturing the benefits of soil biological activity for agricultural production requires adhering to the following ecological principles:

— *Supply of organic matter:* Each type of soil organism occupies a different niche in the web of life and favours a different substrate and nutrient source.

Most soil organisms rely on organic matter for food; thus a rich supply and varied source of organic matter will generally support a wider variety of organisms.

— *Increase of plant varieties:* Crops should be mixed and their spatial-temporal distribution varied, to create a greater diversity of niches and resources that stimulate soil biodiversity. For example, diverse habitats support complex mixes of soil organisms, and through crop rotation or intercropping, it is possible to encourage the presence of a wider variety of organisms, improve nutrient cycling and natural processes of pest and disease control.

— *Protection of soil organisms' habitats:* The activity of soil biodiversity can be stimulated by improving soil living conditions, such as aeration, temperature, moisture, and nutrient quantity and quality. In this regard, reduced soil tillage and minimised compaction - and refraining from chemical use - are of particular note.

Improvement in agricultural sustainability requires, alongside effective water and crop management, the optimal use and management of soil fertility and soil physical properties. Both rely on soil biological processes and soil biodiversity. This calls for the widespread adoption of management practices that enhance soil biological activity and thereby build up long-term soil productivity and health.

Adaptation and further development of soil biodiversity management into sustainable land management practices requires solutions that pay adequate consideration to the synergies between the soil ecosystem and its productive capacity and agro-ecosystem health. One practical example of holistic agricultural management systems that promote and enhance agro-ecosystem health, including biodiversity, biological cycles and soil biological activity is organic agriculture.

Soil Biodiversity

Building of soil fertility is the cornerstone of organic agriculture. Organic practices create suitable conditions for soil biotic and abiotic resources through: manipulation of crop rotations and strip-cropping; green manuring and organic fertilisation; minimum tillage; and avoidance of pesticides and herbicides use. Scientific research in Europe has demonstrated that organically-managed soils significantly increase biological activity and total density and diversity of soil micro-organisms. Such biodiversity enhances nutrients recycling and soil structure. The impact of organic management on soil biological activity and related benefits is summarised below:

— *Abundant earthworms and arthropods:* Organic management increases the abundance and species richness of earthworms and beneficial arthropods living above ground, and thus improves the growth conditions of crops. The biomass of earthworms in organic systems is 30-40 percent higher than in conventional systems, their density even 50-80 percent higher. Compared to the mineral fertiliser system, this difference is even more pronounced. More abundant predators help to control harmful organisms (i.e. pests). In organic systems, the density and abundance of arthropods, as compared to conventional systems, has up to 100 percent more carabids, 60-70 percent more staphylinids and

70-120 percent more spiders. This difference is explained by prey deficiency due to pesticide influence as well as by a richer weed flora in the standing crop that is less dense than in conventional plots. In the presence of field margins and hedges, beneficial arthropods are further enhanced, as these habitats are essential for over-wintering and hibernation.

— *High occurrence of symbionts:* Organic crops profit from root symbioses and are better able to exploit the soil. On average, mycorrhisal colonisation of roots is highest in crops of unfertilised systems, followed by organic systems. Conventional crops have colonisation levels that are 30 percent lower. The most intense mycorrhisal root colonisation is found in grass-clover, followed by the vetch rye intercrop. Roots of winter wheat are scarcely colonised. Even when all soils are inoculated with active micorrhisae, colonisation is enhanced in organic soil. This indicates that, even at an inoculum in surplus, soil nutrients at elevated levels and plant protection suppress symbiosis. This underlines the importance of appropriate living conditions for specific organisms.

— *High occurrence of micro-organisms:* Earthworms work hand-in-hand with fungi, bacteria, and numerous other micro-organisms in soil. In organically managed soils, the activity of these organisms is higher. Micro-organisms in organic soils not only mineralise more actively, but also contribute to the build up of stable soil organic matter. The amount of microbial biomass and decomposition is connected: at high microbial biomass levels, little light fraction material remains undercomposed and vice versa. Thus, nutrients are recycled faster and soil structure is improved.

— *Microbial carbon:* The total mass of micro-organisms in organic systems is 20-40 percent higher than in the conventional system with manure and 60-85 percent higher than in the conventional system without manure. The ratio of microbial carbon to total soil organic carbon is higher in organic systems as compared to conventional systems. The difference is significant at 60 cm depth. Organic management promotes microbial carbon and thus, soil carbon sequestration potential.

— *Enzymes:* Microbes have activities with important functions in the soil system: soil enzymes indicate these functions. The total activity of micro-organisms can be estimated by measuring the activity of a living cell-associated enzyme such as dehydrogenase. This enzyme plays a major role in the respiratory pathway. Proteases in soil, where most organic N is protein, cleave protein compounds. Phosphatases cleave organic phosphorus compounds and thus provide a link between the plant and the stock of organic phosphorus in the soil. Enzyme activity in organic soils is markedly higher than in conventional soils. Microbial biomass and enzyme activities are closely related to soil acidity and soil organic matter content.

— *Wild flora:* Large organic fields feature flora as six times more abundant than conventional fields, including endangered varieties. In organic grassland, the average number of herb species was found to be 25 percent more than in conventional grassland, including some species in decline.

Vegetation structure and plant communities in organic grassland are more even and more typical for a specific site than in conventionally managed systems. In particular, field margin strips of organic farms and semi-natural habitats conserve weed species listed as endangered or at risk of extinction. Animal grazing behaviour or routing activity was found important in enhancing plant species composition. Weeds influence the diversity and abundance of arthropods and flowering weeds are particularly beneficial to pollinators and parasitoids.

— *High-energy efficiency:* Organic agriculture follows the ecosystem theory of a closed nutrient cycle on the farm. Organic land management allows the development of a relatively rich weed-flora as compared to conventional systems. Some "accompanying plants" of a crop are considered useful in organic management. The presence of versatile flora attracts beneficial herbivores and other organisms that improve the nourishment of predatory arthropods. When comparing diversity and the demand of energy for microbial maintenance, it becomes evident that diverse populations need less energy per unit biomass. A diverse microbial population, as present in the organic field plots, diverts a greater part of the available carbon to microbial growth. This increases the turnover of organic matter, with a faster mineralisation and delivery of plant nutrients. Finally, more organic matter is diverted to build up stable soil humus.

— *Erosion control:* Organic soil management improves soil structure by increasing soil activity and thus, reduces erosion risk. Organic matter has a positive effect on the development and stability of soil structure. Silty and loamy soils profit from organic matter by an enhanced aggregate structure. Organic matter is adsorbed to the charged surfaces of clay minerals. The negative charge decreases with increasing particle size. Silt is very susceptible to erosion since it is not charged, but organic-matter layers on the silt surface also favour aggregates with silt.

Genetic Resources for Food and Agriculture

There are several hundred millions of small farmers in the world who do not have the economic means to buy high yielding seeds or the synthetic fertilisers and pesticides necessary for conventional cultivation. Many of these have opted for the maintenance or re-introduction of organic systems based on traditional forms of agriculture. These promote the use of varieties and breeds that are better adapted to local stress conditions and do not require unavailable inputs such as, for example, veterinary drugs.

There are also farmers who have opted for organic agriculture, in part because they wish to produce healthy and environmentally-friendly food, and also because they are attracted by the strong demand for organic products and the related premium prices. Market driven farmers should, as a minimum, rotate crops as the first step towards improving agricultural biodiversity. This is one of the methods required by organic certification bodies as well as by financial programmes. These farmers have also opted for sowing locally-adapted species and varieties that are more resistant to disease and local environmental conditions because synthetic fertilisers and pesticides cannot be relied upon.

The adoption of organic agriculture methods requires farmers to follow a series of agronomic practices that make organically managed systems biologically much more complex than conventionally managed systems. Organic farms make use of larger numbers of plant and animal species than conventional systems. As a result, the large pool of genetic resources for food is maintained and other useful organisms, such as predators, pollinators and soil micro-organisms are increased - for the very benefit of the agricultural system.

The contribution of organic agriculture to the survival of centres of diversity and to under-utilised species, varieties and breeds is described below. The specific requirements of organic farmers for a productive genetic material, in conditions of low-input and harsh environments, are generating innovative approaches to research and development; the emergence of participatory systems of selection and distribution of appropriate genetic material is presented.

Maintenance of Genetic Diversity

The continued cultivation of crop species within their centres of diversity plays a fundamental role in the maintenance of genetic diversity. Preserving the integrity of centres of diversity through ecologically-sound agriculture is an indispensable inheritance for agriculture and as such, for food security for future generations. It is the genetic variability that allows populations to adapt to changing environmental conditions. In centres of diversity, the introduction of organic practices is aiding the conservation, through cultivation, of populations with high genetic variability. The maintenance of agricultural production in centres of diversity requires market outlets for peasants and indigenous communities. The new income opportunities offered by organic markets reverse the present trend of abandoning land which has previously been economically uncompetitive.

Examples where organic agriculture created viable means for *in situ* conservation and use of areas with diverse genetic heritage include: producing and processing cocoa in Mexico to sell Maya chocolate to tourists, and the maintenance of naturally-pigmented cotton in Peru, resistant to pests, diseases and drought. These organic market opportunities provide for the economic viability of thousands of farmers and processors and maintain genetic variability for future generations.

In the past, agriculture has played an important role in the maintenance of genetic diversity. The substitution of a large quantity of species for only a few and the adoption of high yielding and uniform varieties from a genetic point of view, has caused a significant reduction in the genetic inheritance of cultivated species. Many agricultural species, varieties and breeds which have played an important role in the human diet and traditional cultures have practically disappeared over the last century. Organic farmers breed varieties for quality, nutrition, resistance and yield, in reduced input growing conditions.

Research has shown that these characteristics are more likely to be found in older native cultivars. In particular, open pollinated varieties and indigenous breeds offer diverse and regionally adapted characteristics that are better suited to organic agriculture. In the last decade, the adoption of organic agriculture has indirectly established a rescue process of species, varieties and breeds threatened by under-use or extinction. The restoration and enhancement of under-utilised species and varieties has been motivated by specific demands of both consumers and farmers.

For the rescue of varieties threatened by extinction, the development of a market is fundamental and it is here that organic agriculture plays an important role as the price premium gives an additional value to the product. This is especially the case now that there is a consumer' interest in traditional, speciality and organic products. Many consumers search for quality aspects, for either health reasons or culinary traditions. Examples where organic agriculture has allowed the maintenance and improvement of species and varieties that otherwise would suffer strong genetic erosion or extinction include: the discovery of the nutritional value of the gluten-free quinoa in Peru and saraceno grain in Italy; re-introduction of local rice varieties in traditional diets and culture in Indonesia; and economic viability of the Garfagnana spelt in Italy. These cases provided for the survival of poor communities in marginal areas and valorised endangered genetic resources.

Crop Varieties Production under Low-input Conditions

The majority of crop varieties available on the commercial market are not suitable for organic cultivation methods as they have been selected for production dependent on irrigation and large quantities of synthetic fertilisers and pesticides. Many of these are hybrids and are not open-pollinated. In the last few years, the problem has worsened following the arrival on the market of genetically modified varieties.

The selection objectives of organic agriculture differ from those for conventional agriculture. It is of crucial importance to utilise the genotype potential for an improved adaptation of varieties to the low-input conditions prevailing in organic agriculture. The necessity for organic farmers to find methods for obtaining quality products with good yields and limited production costs is greater than for other farmers. Besides the fact that organic farmers cannot apply synthetic inputs, their use of organic fertilisers, natural pesticides and other permitted substances is uneconomical in the long-term. Permitted external inputs as such are relied upon mainly during the conversion period to organic agriculture or under exceptional circumstances. The comparative advantage of certain varieties to withstand local natural stress, especially in marginal areas, leads organic farmers to adopt biodiversity management as their main productive strategy.

Empirical organic breeding systems have been based on the selection of individuals better adapted to the local environment and that are more resistant to pests and diseases. Many of these systems have demonstrated interesting results in restoring and improving local varieties. Examples of restoration of varieties and breeds include: the rescue, in Germany, of an old variety of wheat with a vegetative cycle that allows the absorbency of nitrogen available in sandy soils; in Cuba, the success of local pumpkin varieties is used as the basis for the refinement of methodologies for the selection of varieties for low-input situations.

Demand for breeds to environmental conditions and diseases

Animal breeding for high performance and selecting for early maturity have led to increased susceptibility to diseases, joint inflammation and mastitis as well as circulatory, metabolic and fertility problems of livestock. Loss of breeds is exacerbated by the narrowing genetic base of modern breeds and hybrid lines. The trend towards inbreeding increases the degree of genetic uniformity in the animals and hence, susceptibility to infection, parasites or epidemics.

A significant proportion of local breeds remains in the care of pastoral people and traditional livestock owners in developing countries. Local breeds are suitable for free ranging and robust, thus viable in marginal areas. While the yield may be less in the short-term, animals are more resilient and able to survive in the long-run. Examples where organic agriculture restored, through utilisation, genetic resources resilient to local natural stress include: productive rearing, in Italy, of authochthonous races of the Maremmana cattle, on the verge of extinction, due to its suitability to grow in marshy environments, and the re-establishment of native poultry in South Africa, due to their resistance to Newcastle disease.

Alternative Systems in Organic Genetic Resources

Historically, farmers have managed many varieties and breeds according to agronomic and culinary properties. Considering the need for a wide gene pool to improve and multiply genetic resources for food and agriculture, breeding requires access to seeds and breeds from the formal and informal sectors. Open pollinated varieties, which represent an important gene pool for resource-poor farmers living in marginalised and stress-prone areas, are rapidly vanishing. They are replaced by very few hybrid varieties which require inputs not available to poor farmers and which entail dependence on large seed companies.

Limitations and threats associated with crops have stimulated many organic farmers, especially in the horticulture sector, to produce their own seeds. In order to do this, they have often had to rescue local varieties and develop their own system of selection and distribution. In many cases, the systems include the exchange of seeds between farmers as a fundamental instrument (e.g. organic seed fairs).

Organic systems encourage the preservation and expansion of older, locally bred and indigenous varieties and breeds. Farmers who save their own seeds can gradually increase crop resistance to pests and diseases by breeding for "horizontal resistance". Horizontal resistance is the ability of a crop to resist many or all strains of a particular pest (which differs from breeding for "vertical resistance" to have a gene to resist one specific strain of a disease). By exposing a population of plants to a certain disease or pest, then selecting a group of the most resistant plants and interbreeding them for several generations, a given population becomes more resistant than the original population. Horizontally resistant cultivars are well adapted to the environment in which they were bred, but may be less suitable for other growing conditions.

Benefits derived from new varieties bred by farmers require a legal system of common ownership that allows equitable access and benefit sharing. The biodynamic network of farmers and breeders in Germany provides an example of how such a system could be organised: trials, selection and evaluation of genotypes adapted to low-input conditions is made by farmers and common ownership of new varieties is shared by the community.

Organic agriculture is providing an important contribution to the *in situ* conservation, restoration and maintenance of agricultural biodiversity. The spontaneous establishment of participatory systems of research and development is shaping a simple and practical system of equitable sharing of benefits derived from genetic resources for food and agriculture. The growth pattern shown by the conversion to organic agriculture throughout the world suggests that this contribution is likely to increase still further.

Inter-dependency between Wildlife and Agriculture

Agricultural productivity depends upon the maintenance of ecological balances and the natural properties of plants and animals. The fundamental role of maintaining biodiversity is demonstrated through ecological services such as pollination of crops, predation for biological control of pests, micro-organisms' maintenance of soil fertility and other services vital to the food web. On the other hand, agriculture has the same important role in wildlife conservation, provided that it avoids the use of substances (e.g. pesticides) that could have a harmful effect on natural species and that it maintains food and shelter through a diversified landscape and permanent vegetation (e.g. trees, hedges and fields margins). Finally, a type of land use that provides suitable biological corridors is essential for wildlife conservation.

Nature conservation has traditionally consisted of geographically targeted efforts. While this approach has resulted in a number of successes for rare species or key locations, worrying declines of protected species have occurred. A healthy environment is a prime objective for the conservation of vital terrestrial ecosystems and the wildlife in it. Natural faunal and floral species have strong connections with agriculture, whatever their habitats are, especially as agricultural fields occupy much of the earth's land surface.

Protected areas simply cannot be viewed in isolation from the communities within and near them. Almost everywhere, rural dwellers claim historical rights on protected areas which governments have, at a point in time, declared "protected" for national interest. People inhabiting within or in the neighbourhood of protected areas depend directly on their resources for a living. In India, for example, at least three million people live in protected areas and many other millions live in their proximity.

In Latin America, about 86 percent of national parks are inhabited by indigenous people and migrants. This very dependency on protected areas and its diversity of life forms imposes ecological farming policies. If nature is to be protected successfully, protected area dwellers should be given agricultural choices which are not environmentally destructive and economically rewarding. Considering that the relationship between wild biodiversity and agriculture is reciprocal, the protection of wildlife, biodiversity and natural areas must include a correct management of agricultural systems.

Organic Agriculture and Nature Conservation

There is no doubt that farmers are the most important managers of natural resources. Several studies indicate that organic agriculture safeguards non-agricultural biodiversity and offers a viable alternative in protected area categories where human activities are allowed. Most importantly, the huge land surface surrounding protected areas requires an agro-ecosystem management that preserves the safety and integrity of the landscape. If farm land bordering and connecting protected areas employ organic methods, there is no reason to fear the loss of wildlife or contamination of air, water and soil. These buffer zones are critical to the success of conservation in the protected areas.

Organic agriculture enhances people's ability to live in harmony with nature and to derive economic benefit from their land. Considering that most protected areas traditionally belonged to local villagers, organic agriculture allows local people to maintain some control over their land, protect land and biodiversity through their farming practices, reap its benefits for themselves and, at the same time, conserve and improve the natural environment.

The direct impact that organic agriculture has on ecosystems can be seen on different scales: on-farm, farm margins, and overall ecosystem. While on-farm biodiversity has been discussed in the sections above, the following sections will consider the interactions of organic agriculture with the wider landscape, namely protected areas and buffer zones.

Organic Agriculture in Protected Areas and Buffer Zones

Certain protected area categories allow sustainable land-use activities such as organic agriculture, management of non-timber forest products, fishing, subsistence hunting and ecotourism. Organic farming within protected areas is a growing practice that can "help to define and control sustainable land uses in those protected areas containing significant human populations".

There are some 350 Biosphere Reserves in 85 countries to protect ecosystems. The area surrounding a biosphere reserve, named buffer zone, plays a critical role because activities carried out here strongly influence the core of the protected area itself. Often buffer zones are areas dedicated to agricultural practices. Conversion to organic systems can reduce the detrimental effects of conventional farming, and can provide sustainable systems, suitable for the management restrictions governing buffer zones, and consequently to natural ecosystem conservation.

The protection of the natural heritage must consider the impact, be it positive or negative, that human activities have on it. The characteristics of agriculture make it one of the main activities to be practised in protected areas and buffer zones. Organic agriculture offers a suitable alternative in ecosystems where geographic and morphological conditions are favourable to human activities, such as wetlands and lowland forests. Also, the ecological services offered by organic agriculture in biological corridors are of extreme importance.

Organic Agriculture and Wetlands Conservation

Wetlands are defined in the Ramsar Convention as "areas where water is the primary factor controlling the environment and the associated plant and animal life. They occur where the water table is at or near the surface of the land, or where the land is covered by shallow water". Wetlands are present in every country, from the tundra to the tropics. Their ecological importance derives from their capacity to host high concentrations of birds, mammals, reptiles, amphibians, fish and invertebrate species. In fact, wetlands are boundary areas that combine the components of marine, fluvial and terrestrial ecosystems.

In addition, many functions of wetlands derive from the interactions of different properties of soils, water, plants and animals. Some of these functions include, for example, water purification, water storage, flood mitigation, recharge and discharge of underground aquifers, and stabilisation of local climatic conditions. Natural wetlands are among the most threatened ecosystems in the world. Their high productivity and the morphologic characteristics make wetlands excellent areas for many human activities, especially agriculture. Often the damage caused by land reclamation for agriculture or unsustainable practices leads to the disappearance of wetland areas and corresponding biodiversity. Suitable management is therefore a critical priority to save these fragile ecosystems.

Organic agriculture can help wetland conservation, by providing suitable habitats for wildlife species, reducing water pollution and, at the same time, offering a valid economic alternative to the exploitation of natural resources. Examples where the avoidance of synthetic inputs and cropping strategies employed by organic farmers in wetlands offered breeding and feeding habitats for endangered wildlife include: return of cranes and storks in the cereal production in Muravia Park, Russia and of egret and heron in rice production in the El Ebro Delta, Spain. Organic beef production in the Pantanal Region, Brazil, created natural grasslands vital to wild mammal herbivores in an area considered the world's biggest ecological sanctuary and which previously suffered from deforestation of savannah and implementation of artificial pastures for beef.

Organic agriculture in protected forest areas

Human activities such as modern agriculture and grazing can be a serious threat to forest ecosystems and hence, for a large percentage of the world's flora and fauna. In fact, the practice of clearing the tree cover often interrupts the continuity of canopy, an essential characteristic for genetic and specific biodiversity flow. In many cases, the same agricultural fields are also a barrier to wildlife movement, and pollution from agrochemical abuse may have negative impacts on forests.

Organic management in forest areas has the potential to reduce the loss of biodiversity caused by these agricultural activities. The organic systems utilised in forest areas are shade and sun cultivations. In the first case, the plants are grown under a canopy that consists of original forest trees or selected plants. In the latter, there is no forest cover on the agricultural field. Although yields of full sun production are often higher than for shade cultivations, the latter provide timber, fuelwood and other fruits, and have a lower soil erosion rate. Shade plantations have been shown to be highly beneficial to biodiversity conservation in tropical forest ecosystems, including millions of migratory birds and other animals and plants.

For example, the Smithsonian Migratory Bird Centre's research has demonstrated that bird species are almost twice as large in shade coffee as in sun coffee. Shade coffee cultivation therefore offers important conservation opportunities as the structural profile of shade coffee farms is similar to natural forest, providing habitat suitable for resident birds and migrants. As a result many scientists and conservationists believe that shade coffee plantations are ideal zones for migratory birds. The Northwest Shade Coffee Campaign has obtained statistical data on shade coffee as compared to sun coffee and has shown how shade cultivations are preferred by bird species, mid-size mammals and several species of ants, beetles, amphibians, epiphytes and others. Smithsonian Migratory Bird Centre's studies indicates that at least 180 species of birds live in Mexican shade coffee and cocoa fields, and that 90 percent fewer bird species live in Colombian sun coffee plantations than in shade coffee.

Researchers highlighted the capacity of the canopy cover to support secondary structures (e.g. epiphytes, parasites, mosses and lichens), which in turn support arthropods, amphibians and other living beings. Another relevant fact is that shade coffee hosts a large density and diversity of predators and parasitoids involved in the control of insect pests. Shade coffee offers an optimal habitat also for other tropical forest species, such as beetles, ants, wasps and spiders, and supports a high diversity of many vertebrate groups. Finally, areas of high ecological value located around coffee farms are protected.

Examples of the co-existence of shade organic cultivations and richness of biodiversity in forest ecosystems include: shade coffee in the buffer zones of El Trionfo Biosphere Reserve, Mexico; shade cacao in the tropical forest of Montes Azules Biosphere Reserve, Mexico; and yerba mate in the threatened Atlantic rainforest of the Guayaki Biological Reserve, Paraguay.

Examples of conversion to organic agriculture to reduce pressure on endangered forests include: land exchange between Del Oro orange production and Guanacaste Conservation Area in Costa Rica to restore rare native forests; organic bananas to protect the threatened Guaraquecaba Atlantic forest and its extraordinary biodiversity in Brazil; and the re-introduction of native species through agroforestry in Ampay Forest Sanctuary in Peru. In all these cases, local communities' income was raised while providing benefits to the environment.

Organic agriculture in biological corridors

Corridor zones link protected areas with one another and either remains under wild cover, or are managed to ensure that human land use is compatible with the maintenance of a high degree of biological connectivity. These areas have a fundamental ecosystem function forming part of a large-scale ecological web. To be effective, biological corridors must offer suitable habitats to wildlife, thus the maintenance of a healthy environment is necessary. Areas joining parks or reserves are equally important for biodiversity conservation as they ensure ecosystem connectivity.

These "linkage" areas often host agricultural systems but if they are managed in an unsustainable manner, the vitality of these corridors decreases or is totally hindered. Careful organic management in agricultural land between the protected areas has permitted the creation of important biological corridors and allowed the protection and increase of biological diversity. Through on-farm structured vegetation and canopy tree diversity, organic agriculture has demonstrated its capacity to provide livelihoods to farmers while providing easier movement of animals between managed forests and protected areas.

One of the largest, most unique conservation efforts presently underway in the world is the Meso-American Biological Corridor that runs across seven countries, covering an area of 770 000 km^2, from Panama to Mexico. Examples of organic cultivations include: shade coffee in the buffer zone between El Imposible and Los Volcanes National Parks in El Salvador where hundreds of endemic and threatened birds, mammals, reptiles and trees strive; and shade cacao and bananas in the Talamanca-Caribbean Biological Corridor and its buffer zones, Costa Rica, where most of the fauna is endemic.

Organic systems offer habitats similar to intact forests, providing suitable refuge for migratory and resident birds and foraging for many forest mammals, including species at risk. Organic land management and bird conservation programmes exist in several northern countries. Examples include: organic crop and livestock production in Brandenburg, Germany where the higher presence of skylarks has been observed; and organic farming inside a wildlife corridor that connects Adirondak National Park and Lake Champlain in the State of New York, USA.

8

Sustainable Agriculture

Various components in the conservation agriculture practices could be identified and also classified between those done in the beginning, then the second level of transition practices and the final stage practices. Whiles adoption of conservation agriculture is noted in all the main household categories, closer look at these trends have indicated that though all are adopting, they do so for different reasons. Some of these reasons are underlined by the socio-economic and traditional category and disposition of the household. This highlights an important factor in understanding adoption or none adoption decisions.

Conservation Agriculture (CA) is being understood as a system of farming that conserves, improves and makes more efficient use of natural resources through integrated management of available soil water and biological resources. The fundamental principle underlining CA practices is to achieve sustainable soil-water productivity through crop rotation, reduced disturbance to the soil structure, protection of the soil from direct climatic impacts such as solar radiation, rain and wind, enhance water infiltration and build-up of soil organic matter and soil organisms.

Whiles for the poorer households it would be survival strategies with higher risks, but also given in to fate with a feeling of "nothing to lose", for the more financially able it would be adventurism or simply demonstrating that financial or social mighty. This also goes to explain why the poorer households expected to be at margins of survival appear to go ahead in adoption. This may also relate to the fact that the poorer households who are also aware of their difficulties have one of the key benchmarks in starting to look for a solution—noting and accepting that there is a problem. Some development efforts have passed very unsuccessfully because after noting a critical problem in a community have gone on to provide solutions without taking time to ensure that the local community felt the same about the issue as a priority problem that needed solutions.

Changes in Inter-farm Linkages

The Factor	*Noted changes in inter-farm linkages induced by the practice*	*Reasons for these changes*
Farm labour and labour use	Households involved with CA have indicated a wider spreading of labour requirements through the year, i.e. over a longer period. As members of the groups, it has also become possible that the group spend time working in one member's field. This way, especially the very old and disadvantaged have had the work done and done in time.	Members of the group seeing and taking more collectively responsibility and with the application of CA all have some time they can spare for such collective activities.
Perception and attitude on	Group members and the community in	Community was becoming aware that

Contd. . .

communal resources such as communal grazing land, roads, water points and streams	general are coming from a background were community development and care/ownership for all community resources was a responsibility of government. The CA-FFS approach has made many community members take ownership and responsibility for community property, including this like public roads.	the degradation in their field could not be controlled unless the effort was collective and done even in neighbouring fields.
sharing of knowledge	Members of the group now sit and opening share ideas and experiences on their farming. In the past this was equal to helping or giving knowledge to a competitor. Many of the Eutolelo group members have also gone as resource persons (informal) to other villages and willingly explained CA	
Livestock ownership and grazing patters	Very significant changes have happen in this regard:- - Village now has widely accepted by-laws governing communal livestock issues. It is now easy to deal with cases of livestock damaging other people's crop. - To avoid problems livestock owner have made sure that there is good care for the livestock.	- Heavy punishment for offenders - General positive appreciation for restricted livestock movements - Farmers/community aware of the benefits of letting trees grow and soil not disturbed.

Contd. . .

	Have done this by reducing the number, growing of fodder for supplementary feeding, etc... - Moved the livestock to far away holding plains	
Finances and household access to finances	There is increase in disposable income in an average household; but even more important has been the stability in this income source and the fact that to a certain extent one is more sure of that income. Through group savings a household also has access to possible finances they can use for investment or in difficulty times.	
Richer-poor relationships and linkages	The group have allowed everyone regardless of social status and this has improved relationship across social classes. Even though for different reasons, all seem to be well adjusting in adoption of CA. There is also more positive support for those unable, e.g. old couples - positive in the sense that its not any more like getting ride of excess but human desire to help.	
Physical changes	Reduced erosion has been seen and felt across the fields and plots. Gullies, which were running from field to field and across footpaths and road are being converted into productive land.	More soil cover and increased rainwater harvested into locations it is falling hence reduced flow.

Local Ownership

The application of conservation agriculture has been noted as in many aspects very well within local beliefs and practices. It is generally believed in the area that use of inorganic fertilisers eventually makes the soil infertile and that there will be no yield from that field unless the fertilise (inorganic) is used. The local people believe fertiliser is organic (kraal or composite) manure. This is compatible with conservation agriculture, which while not discouraging use of inorganic fertilisers acknowledges that organic fertiliser is essential in sustaining both the physical and chemical fertility of the soil. What Conservation agriculture has done is to explain some things farmers have been aware of such as declining productivity in a case of extension use of mineral fertilisers and no organic fertiliser.

Only making planting holes and place seed was an acceptable practice once upon a time. However until the advent of conservation agriculture such a practice came to be regarded as unacceptable and only for poor households. It was also a sign of disorganised household unable to plan and do the ploughing in time—it was only done when you have failed to plough. Conservation agriculture has brought credibility to this practice meaning that even those without much labour could work and crop sizable portions of cropland.

Conservation agriculture has not introduced any typically new crop but allowed new uses and more production for crops that could have been grown only as backyard garden crops. The fact that farmers have been able to select and start from different aspects of CA has enabled them see CA as "friendly" and adaptable. Some farmers start with reduced tillage, i.e. basins or ripping while others start rotations and intercropping. In almost all cases, farmers have started adoption on a smaller portion of their land and expanding year after year. With regard to inputs, one critical input/item required that is sometimes a constraint is appropriate equipment for those using draft animals, let alone tractors. Access to appropriate conservation agriculture equipment is usually a big problem. Some of these equipment cannot even be found in shops even in big centres like Dar-es-salaam.

Environmental Practices

Soils

Putting literally, the farmers applying CA were quoted as say, "now we have soil". This is in reference to the fact that in conventional practices cropping fields are cleared bare of all the top soil—remaining exposed is the hard, rocky infertile sub-soil. With conservation agriculture—both water and soil held within the field, both kraal manure and crop residue going towards organic matter replenishment farmers have observed rebuilding of soil including some life activity (worms and microorganisms) in the soil. With increased soil activity, especially water retained and available for plant use on a much longer period, farmers are also able to grow more and even in terms of the resulting total biomass (weeds included). Though with training and awareness and access to external inputs some households are getting to also use some mineral fertilisers, some good yield (in quantity and in stability) has been observed even with manure use only.

Water Quality and Quantity

Let alone quality, the application of conservation agriculture has simply made water much easily available in many households—both for domestic use such as washing clothes and utensils and for their livestock. Normally one had to walk long distances up to 8 km to fetch water. This impact on water availability has gone beyond just the practicing households, as even non-practicing neighbouring households are able to utilise the water from the collecting ponds.

Fostering Factors

Initially many farmers were not able to cope with the challenges of adopting conservation agriculture. Reasons included mere reluctance to change/take up the risks, rainfall uncertainties and social problems such as destruction of trees, live cover crop and contours by animals. Lack of implements was another key limitation for many households. The matter was worse and more complicated in cases when the local agriculture extension staff were themselves not knowledgeable or unconvinced about conservation agriculture.

Other key challenges that at some instances discouraged or hindered adoption included:

- — Weed management which appeared to be more demanding with reduced soil disturbance,
- — Availability of seed and other planting materials for new crops especially cover crop seed.
- — Appreciating the long term implication in CA adoption, farmers were unwilling to apply the same on borrowed land.
- — Farmers with very small portions of land (an acre or less) had difficulty introducing rotations

Economic Viability

The group income generating activities with conservation agriculture based field work have regenerated the cash economy in the village. Crop production is now said to provide more viable business and significantly higher in terms amount of money brought in. Whiles livestock would provide manure and security for quick cash money needs, CA based crop production is said to be growing in reliability and in amount of cash realised.

Financial Viability

Compared to costs the family has to incur, the system comes out as highly profitable even when a cost is put on the family labour. Main costs incurred are on seed and fertilise and sometimes and in some cases on hired labour. This is also because the systems have been adapted on intensive use of locally available resources whose main cost is the time and energy used in collecting, transporting and applying/using. Therefore, simple calculation a household would retain upto 60–70 % profitability. It should also be noted that these households are using very little if at all of inorganic fertilisers.

Social and Cultural Acceptance

Local Traditions and Belief Systems

Part of the main challenges to the adoption of conservation agriculture has been compatibility with socially accepted and widely practiced norms. This is also the area that the group could be noted as achieved greatest non-technical impact—Impact just as important as the technical conservation agriculture application. The main social challenges can be distinguished as:

i) self organisation into groups that takes collectively responsibility for community development

ii) accepted livestock keeping practices also noting that the community is coming from a once nomadic 100% pastoralists tradition

iii) others

The community, though with a lot of extended family connections and an accepted political village government in place, the village still had great difficulties in managing issues that went beyond the household. A case closer to this subject matter was the management of livestock which everyone agreed the solution had to be at community level but could not get round to any workable mechanism. With all the attributes of owning livestock in a household, the village was aware of the consequent destruct impacts from main the free loaming and uncontrolled grazing livestock keeping patterns. The livestock destroyed contours, were damaging the trees and the trampling was only loosening the soil for easy erosion.

The village government, expected to be an authority on community matters in the village, reacted very slowly to reports of one household's livestock destroying another's crop or contours, etc... And when the action came it was often no more that a weak advise to restrict ones animals. With the build up and appreciation of group approaches, the village has been able to come together and agree on communally respected norms and standards for livestock management. The community has even gone further to put in place mechanism for punishing the offenders—mainly based on providing compensation to the offended. The offending also pays a fee to the village government, which goes into the community bank account.

Other related social, cultural and traditional norms include burning of crop residue widely practiced and accepted as part of land preparation. This is significantly changing both in attitude and in practice—apart from penalties given to household originating a bush fire, the change could also be attributed to the fact that the community is seeing the immediate positive impact with regard to reduced dust (wind erosion), less water erosion.

Application of conservation agriculture also demanded more open and direct interaction between women and men at community level and husband and wife in a household. To an extent this also implied cutting across accepted gender classification in duties and responsibilities. Instead of women taking responsibility for household food crop production and men responsible for the cash field crop, CA demands integration of the two compelling more women-men interaction with both men and women jointly taking up the once gender classified tasks such as land preparation and weeding .

9

Agricultural Biodiversity

Agricultural biodiversity is a vital sub-set of biodiversity. It is a creation of humankind whose food and livelihood security depend on the sustained management of those diverse biological resources that are important for food and agriculture. Agricultural biodiversity, also known as agrobiodiversity or the genetic resources for food and agriculture, includes:

— Harvested crop varieties, livestock breeds, fish species and non- domesticated ('wild') resources within field, forest, rangeland and in aquatic ecosystems;

— Non-harvested species within production ecosystems that support food provision, including soil micro-biota, pollinators and so on; and

— Non-harvested species in the wider environment that support food production ecosystems (agricultural, pastoral, forest and aquatic ecosystems).

Agricultural biodiversity results from the interaction between the environment, genetic resources and the management systems and practices used by culturally diverse peoples resulting in the different ways land and water resources are used for production. It thus encompasses the variety and variability of animals, plants and micro-organisms which are necessary to sustain key functions of the agro-ecosystem, its structure and processes for, and in support of, food production and food security.

Plant Genetic Diversity in Crops

The plant genetic diversity used in agriculture—the crops that feed us and their wild relatives—is being lost at an alarming rate. Just nine crops (wheat, rice, maize, barley, sorghum/millet, potato, sweet potato/yam, sugar cane and soybean) account for over 75 percent of the plant kingdom's contribution to human dietary energy. None of the world's staple crops is likely to disappear. Yet they, too, are threatened—not by the loss of a single crop species such as wheat or rice, but by the loss of diversity within species.

All major food crops, the staple crops grown and consumed by the vast majority of the world's population, have their origins in the tropics and subtropics of Asia, Africa and Latin America. Over the years, farmers selected and domesticated all major food crops on which humankind depends today. Wheat and barley originated in the Near East, for example. Soybeans and rice came from China. Sorghum, yams and coffee came from Africa. Potatoes and tomatoes originated in the Andes of South America, and maize in South and Central America.

Crop genetic diversity is still concentrated mainly in regions known as "centres of diversity", and located in the developing world. Farmers in these areas, who still practice traditional agriculture, cultivate local varieties known as "land races" that have been selected over many generations. Closely related species that survive in the wild are known as "wild relatives" of crops. Together, land races and their wild relatives are the richest repositories of crop genetic diversity.

Thousands of different and genetically distinct varieties of major food crops owe their existence to millions of years of evolution and to careful selection and nurturing by our farmer ancestors during some 12 000 years of agriculture. This diversity protects the crop and helps it meet the demands of different environments and human needs. Potatoes, for instance, originated in the Andes, but nowadays they can be found growing below sea level behind Dutch dykes or high in the Himalayan mountains. One variety of rice survives on just 60 centimetres of annual rainfall, another floats in 7.5 metres of water.

FAO estimates that since the beginning of this century about 75 percent of the genetic diversity of agricultural crops has been lost. We are becoming increasingly dependent on fewer and fewer crop varieties and, as a result, a rapidly diminishing gene pool. The primary reason is that commercial, uniform varieties are replacing traditional ones—even, and most threateningly, in the centres of diversity. When farmers abandon native land races to plant new varieties, the traditional ones die out.

The introduction, beginning in the 1950s, of high-yielding grains developed by international crop breeding institutions led to the Green Revolution. The spread of the new varieties in the developing world was dramatic. By 1990 they covered half of all wheat lands, and more than half of all rice lands—a total of some 115 million ha. This resulted in large increases in yields...but large decreases in crop diversity.

Megacentres of Cultivated Plants

The erosion of crop genetic diversity poses a serious threat to food supplies. To maintain pest and disease resistance in major food crops, for instance, or to develop desirable traits such as drought tolerance or improved flavour, plant breeders require fresh infusions of genes from the farms, forests and fields of the developing world. Developing the high-yielding, élite cultivars of modern agriculture depends on a steady stream of new, exotic germplasm. Plant breeders continuously try to develop new varieties to keep one step ahead of thousands of pests and diseases. Without access to traditional land races and their wild relatives, modern agriculture would be seriously endangered.

Genetic Uniformity

Industrialised agriculture favours genetic uniformity. Typically, vast areas are planted to a single, high-yielding variety—a practice known as monoculture—using expensive inputs such as irrigation, fertiliser and pesticides to maximise production. In the process, not only traditional crop varieties, but long-established farming ecosystems are obliterated. Genetic uniformity invites disaster because it makes a crop vulnerable to attack—a pest or disease that strikes one plant quickly spreads throughout the crop.

The Irish Potato Famine of the 1840s is a dramatic example of the dangers of genetic uniformity. None of the few varieties of the New World potato introduced into Europe in the 1500s were resistant to a potato blight that struck Ireland in the 1840s. The potato crop was wiped out. Over a million people died in the famine and a million more emigrated to the New World. More recently, in 1970, genetic uniformity left the United States maize crop vulnerable to a blight that destroyed almost $1 000 million worth of maize and reduced yields by as much as 50 percent. Over 80 percent of the commercial maize varieties grown in the United States at that time were susceptible to the virulent disease, southern leaf blight. Resistance to the blight was eventually found in an African maize variety called Mayorbella. A major catastrophe was averted by incorporating this resistance into commercial varieties.

Crop Genetic Diversity in Modern Plant Breeding

The value of genetic diversity to modern plant breeding is enormous. The United States Government estimates that a 1 percent gain in crop productivity means a $1 000 million benefit to the American economy. Italian scientists calculate that the benefits of exotic germplasm for a single crop, durum wheat, amount to $300 million per year. Not only cultivated species but also the genes from wild relatives are enormously valuable. Between 1976 and 1980, wild species contributed an estimated $340 million per year in yield and disease resistance to the farm economy of the United States.

Stunted Rice

During the 1970s the grassy-stunt virus devastated rice fields from India to Indonesia, endangering the world's single most important food crop. After a four-year search which screened over 17 000 cultivated and wild rice samples, disease resistance was found. Only one population of the species *Oryza nivara*, growing wild near Gonda in Uttar Pradesh, was found to have a single gene for resistance to grassy-stunt virus strain 1. Today, resistant rice hybrids containing the wild Indian gene are grown across 110 000 km^2 of Asian rice fields.

In the developing world, crop genetic diversity enables farmers to select crops suited to ecological needs and cultural traditions. Without this diversity, options for long-term sustainability are lost. This is particularly true in marginal areas with highly varied environments. The variety to a large extent determines the need for fertilisers, pesticides and irrigation. Communities that lose traditional varieties, adapted to local needs and conditions over centuries, risk becoming dependent on external sources of seeds and the inputs needed to grow and protect them. Without an agricultural system in harmony with a community and its environment, self-reliance in agriculture is impossible.

To feed an increasing world population, all available genetic resources, including wild relatives, will need to be tapped. Modern plant breeding as well as new biotechnologies offer the potential to exploit little-known plant species as sources of food, and to enhance the qualities of those plants that are underutilised—especially traditional plants of special significance to poor people, such as local grains, legumes, oilseeds, fruits and vegetables.

Traditional food crops, often grown by rural families to see them through the "hungry season" just prior to harvest, offer many advantages. Many of them are drought resistant, can be grown without expensive inputs and have good storage qualities. For many developing nations, self-reliance in food production will depend on low-input agriculture in poor production environments. The capacity to grow varieties, particularly those resistant to pests and diseases and adapted to marginal lands, is vital for sustainable agriculture and food security.

Geopolitics of Plant Genetic Resources

Historically, scientists from the industrialised countries have ventured southwards in search of exotic plants for plant breeding. Seeds found in tropical centres of diversity have been freely collected and later deployed in plant breeding. As a result, much of the collected diversity of Third World origin has come to be stored in the northern hemisphere or in gene banks established by developed countries.

The issue of control, ownership and access to plant genetic diversity has come to the fore over the past two decades. Plant breeding in the industrialised world has become increasingly commercialised and is now dominated by transnational seed and agrochemical corporations. To promote innovation and to enable breeders to recoup their research investment, many governments in the industrialised world have adopted a system of "plant breeders' rights". This gives patent-like protection to breeders with limited monopoly rights over the production, marketing and sale of their varieties for a period of up to 20 years.

The disparity between unrestricted access to genetic resources, including farmers' land races, and the existence of proprietary rights such as "breeders' rights" on improved varieties has fuelled intense debate over the inequity in the flow of germplasm from the developing world to the industrialised world. At the United Nations, representatives from the developing world ask: Why are patented seeds, originally from developing countries, bringing profits to seed companies in the industrialised countries without corresponding compensation for the developing world? What compensation will be made to those who have tended and nurtured the world's crop genetic diversity and continue to conserve and make it available today?

Use and Conservation of Plant Genetic Resources

The farmer uses plant genetic resources as seeds or vegetatively propagated material; they are often the one input that farmers can produce for themselves. FAO assistance includes projects for the production and use of good-quality seed, training and guidance in propagation and multiplication, quality control, and processing, storage and distribution of improved seed. FAO provides samples and information to research centres, scientists and field projects for use in crop introduction, evaluation and breeding.

Several thousand plant species have been used for human food in history, but now only about 150 are cultivated and no more than three supply almost 60 percent of the calories and protein derived from plants. Since the beginning of this century about 75 percent of the genetic diversity of agricultural crops has been lost. Kenaf (*Hibiscus cannabinus*), an East African plant related to cotton and okra, may provide an alternative source of pulp for making paper; in the southern United States it yields three to five times more pulp than trees do and requires minor chemical treatment to whiten the fibres.

From wild pineapples found in the dry open Chaco of South America, breeders have imparted high-sugar content and a distinctive "wild fruit" flavour to cultivated varieties. Genes transferred from a wild relative of the tomato found on the shores of the Galapagos Islands has conferred salt tolerance to cultivated varieties so that they can be irrigated by one-third sea water.

FAO, as a sponsor of the Consultative Group on International Agricultural Research (CGIAR), supports plant breeding and other research carried out at international research centres. Many of its projects focus on traditional food crops such as roots and tubers which in some developing regions contribute up to 46 percent of total calories consumed.

Roots and tubers can tolerate a wide range of conditions and are well suited to traditional farming systems. They can be intercropped with other plants and most of them can be grown year-round, providing extra calories during the hungry season. Traditional crops have yet to be explored genetically, but their potential for improvement through breeding seems promising.

Northern Europe are usually unable to sustain high production in less hospitable environments. They require intensive management and high levels of inputs such as high-protein feed, medication and protective housing. Introduction of intensive animal production creates dependency on imported technologies: it is neither affordable nor sustainable for most farmers in the developing world.

After thousands of generations of controlled interbreeding, most domesticated animals no longer have wild relatives from whom germplasm can be obtained. When a variety becomes extinct, an already narrow genetic base shrinks irreversibly. Commercial breeds suited to intensive production do not offer an adequate genetic reservoir for the future. Their genetic base reflects the emphasis on maximising production. The turkey that is mass-produced on factory farms in North America and Europe, for example, has been selected for such a meaty breast that it can no longer breed unassisted. This broad-breasted breed —which accounts for 99 percent of all turkeys in the United States today— would become extinct in one generation without human assistance in the form of artificial insemination.

Animal Genetic Diversity

The genetic diversity now found in domestic animal breeds allows farmers to select stocks or develop new breeds in response to changes in the environment, threats of disease, market conditions and societal needs, all of which are largely unpredictable. Indigenous livestock breeds often possess valuable traits such as disease resistance, high fertility, good maternal qualities, longevity and adaptation to harsh conditions and poor-quality feed, all desirable qualities for low-input, sustainable agriculture.

The rare Taihu pigs of China, for instance, offer valuable traits for swine breeders worldwide. This group of pigs has thick, wrinkled skin and long, droopy ears. They can use a high proportion of forage foods in their diet. The adult pig has little lean meat—whence the Chinese passion for sucking pig. But Taihu pigs reach sexual maturity in just 64 days and are extraordinarily fertile, producing an average litter of 16 piglets compared with only ten for western breeds. Researchers in Europe and the United States are exploring ways to incorporate these beneficial qualities into commercial breeds. A company in the United Kingdom, National Pig Development, has already produced a commercial hybrid of the Meishan, one of seven strains of Taihu pig. Announced in 1992, it combines the fecundity of the traditional Chinese breed with a higher lean meat content.

Thirty per cent of Africa's cattle population, approximately 160 million cattle, are at risk from trypanosomiasis — a debilitating and frequently fatal disease transmitted by the tsetse fly in 36 African countries covering over 10 million square kilometres. This devastating disease jeopardises not only African milk and meat supplies, but important by-products and services such as hides, manure, fuel and draught power. Annual losses in meat production alone are estimated at US$5 000 million.

African cattle

Several traditional African cattle breeds, among them the small humpless N'Dama, have developed resistance (trypanotolerance) over thousands of years of exposure to the parasite — a trait that relatively modern African breeds do not possess. This genetically based resistance offers hope of reducing or controlling the impact of trypanosomiasis. Small numbers of trypanotolerant N'Dama cattle have long been maintained by West African farmers in marginal farming areas. They thrive on low-quality forage and, though less productive than modern breeds of cattle, their high survival and reproductive rates and longevity make them extremely valuable in harsh environments.

Using a technique known as "embryo transfer", the population of trypanotolerant N'Dama cattle has already been increased in order to conserve this rare breed, improve its performance and study its disease resistance. The N'Dama's hardiness, heat tolerance and disease resistance have also been recognised. N'Dama cattle have been crossed with the Red Poll, an endangered British breed, to produce the Senepol breed. The Senepol has been introduced successfully in the Caribbean and the southern United States.

Conserving Animal Genetic Diversity

There is already less genetic diversity in farm animals

than in crop plant species and over a third of the remaining animal genetic resources is currently at risk. In 1992, FAO launched a comprehensive programme for the global conservation of animal genetic resources. It includes:

— A global inventory of animal genetic resources including a database to characterise and enumerate all breeds of livestock used in agriculture.
— Action to identify breeds at risk of extinction as well as ways of protecting them.
— Promotion of programmes in developing countries to conserve endangered breeds in their native habitats. The aim is to enhance the attraction of indigenous breeds at risk of being substituted by imported breeds which are often brought in without considering local conditions or sustainability.
— Improvement of livestock breeding capacities in the developing world. In particular, new technologies will be used to identify livestock diversity and the specific genes responsible for valuable traits.

FAO is exploring the possibility of establishing a global centre for domestic animal genetic diversity to serve as the focus for efforts to overcome the present erosion of these irreplaceable resources and to promote their effective and sustained use. Conservation of animal genetic diversity is essential to global food security and to protect our ability to meet the challenges of the future.

Fish and Aquatic Biodiversity

Oceans, lakes and rivers cover four-fifths of the earth's surface, but little is known about their living resources.

Fewer aquatic than terrestrial species have been described, but there is no reason why aquatic biodiversity should be less.

Tropical waters are the richest in terms of species diversity. The Indo-West Pacific Ocean, for example, contains an estimated 1 500 species of fish and over 6 000 species of mollusc, compared with only 280 fish and 500 mollusc species in the Eastern Atlantic. Inland waters are also rich in diversity, the greatest concentration once again being in the tropics. Thailand, for example, could have as many as 1 000 species of freshwater fish, but so far only 475 have been documented. Brazil is believed to have more than 3 000 freshwater fish species — three times more than any other country.

For the most part, the aquatic harvest consists of wild rather than farmed species. World production, 90 percent of it finfish, stands at almost 100 million tonnes a year. Of this, only about 13 million tonnes come from aquaculture. Over 4 million tonnes of algae are also harvested annually.

Fishing, fish processing and fish trading have provided food, employment and income in coastal and inland communities for centuries. Fish contribute substantially to the world supply of animal protein, either directly or through their use as feedstuff for livestock — almost a third of the fish catch is converted into meal and oil.

The developing countries account for more than half the world catch. Their fisheries are dominated by small-scale or artisanal producers. Artisanal fisheries, typically using small boats and canoes, account for more than 25 percent of the world catch. They supply more than 40 percent of the fish used for human consumption. These fisheries are also a significant source of employment —

an estimated 100 million people in the developing world depend upon them for all or part of their livelihood.

By the turn of the century, demand for fish is expected to exceed by some 20 million tonnes the productive capacity, estimated at about 100 million tonnes, of stocks now exploited by the capture fisheries. Increased incomes and appreciation of the dietary value of fish are spurring the demand for fish and fish products in the industrialised countries, especially for luxury products such as oysters, shrimp, salmon and tuna. In the developing regions, population increases and the need to tap every potential source of food and foreign exchange provides the main impetus for increased fishing activities.

One response to the growing demand for fish and its falling availability has been the development of aquaculture. This rapidly expanding source of food poses some threats to biodiversity by concentrating on a very small range of species and an equally narrow genetic base in these species. Large-scale escapes of cultured fish, or deliberate releases of stocks for ranching, are thought to influence the genetic composition of the wild resource.

Troubled Waters

Aquatic biodiversity is threatened primarily by human abuse and mismanagement of both the living resources and the ecosystems that support them. Loss of habitats, over-exploitation and introduction of exotic species are the prime hazards.

Overexploitation: Fish stocks are a renewable resource, but already many of them are strained to the limit. Over the years, they have suffered from a widespread notion that the seas are inexhaustible, economic pressures that have encouraged overexploitation and, until just over a decade ago, an international regime that gave almost unlimited access to the majority of them. All fishing activities depend on a fragile resource base which, if mismanaged and overexploited, can easily collapse.

Efforts to regulate marine fisheries can be traced back to the late 1800s with the creation in Europe of the Intergovernmental Commission for the Exploration of the Seas (ICES). Many fishery bodies for developing and regulating fisheries, in both marine and inland waters, have been established since — nine of them under the auspices of FAO. Despite this appreciation of the threat posed by overfishing, stocks have continued to be exploited at a non-renewable rate.

All demersal (deep water) species such as cod, haddock and pollack are now either fully exploited, overfished or depleted. Larger pelagic (surface water) species such as herring, sardines and anchovy, stocks of which can fluctuate greatly from year to year, are in serious need of management. Crustacea such as shrimp, lobster and crab are also overexploited. Only the bivalve molluscs, such as mussels and clams, and cephalopods such as squid and octopus, offer much scope for expanded production.

The world fish catch has increased more than fourfold in the past 40 years, but the misuse of modern technology, coupled with government support for otherwise non-economic production, has had a devastating impact on fish stocks. Fleets using sophisticated fish detection, non-selective nets (up to 50 km long) and bottom trawls are driving some species to extinction. FAO estimates that the cost of overexploitation amounts to some US$30 000 million per year.

The impact of overexploitation of fisheries may be greatest in the developing world. Commercial fishing in tropical waters can often mean valuable foreign exchange for developing nations, but it can also lead to intense competition with declining catch rates for small-scale fisheries, many of which provide fish for local consumers and markets. Higher fish prices, the result of increased demand exacerbated by overfishing, are making fish unaffordable to an increasing number of poor people. Fish is no longer "a cheap meat dish" - a marketing slogan used in the United Kingdom in the 1950s.

Selectivity of Fishing Methods

Traditional fishing gears, ranging from a simple harpoon to a basket- work fishtrap, are typically selective for both size and species and are adapted to the diversity of fish captured, whereas commercial gears, such as the purse seine, large driftnet and trawl, often have a by- catch of unwanted species. The displacement of traditional fishing methods, combined with the introduction of new materials and highly mechanised fisheries, has contributed to overexploitation of resources in both marine and freshwater environments.

Environmental degradation: To the pressure of exploitation must be added the degradation or destruction of aquatic ecosystems caused by pollution or competing uses. The oceans function as a sink for carbon dioxide, eroded soils, contaminants, fertilisers, human and industrial wastes. Most urban and industrial activities and, indeed, much of human life, are concentrated close to coastal waters, rivers and lakes. Six out of ten people live in coastal areas, and migration towards them is increasing.

The development of intensive aquaculture has, in some cases, damaged coastal ecosystems and water resources, causing conflicts over land use and resources, and even undermining local sources of employment and food. In parts of Asia, thousands of hectares of rice paddy have been replaced by high-value shrimp farming or had their productivity reduced by salinisation caused by neighbouring aquaculture enterprises. In the Indo-Pacific, more than one million hectares of mangrove forests have been converted to aquaculture ponds. Mangroves provide spawning and nursery areas for many marine species and are vital to maintaining ecological balance and biodiversity.

Exotic Species

The introduction of exotic fish species can have many unforeseen consequences. The release of the Nile perch in Africa's Lake Victoria is a classic example. Introduced in the late 1950s as a sports fish, its voracity and large size has driven many of the smaller indigenous species to extinction. Some scientists speculate that 200–300 species of fish may have been lost.

The expanding population of Nile perch is making Lake Victoria one of the most productive lake fisheries in the world, yielding 200 000 to 300 000 tonnes per year. But increased productivity may have been achieved at serious ecological and social cost. The lake is increasingly providing fish for export rather than local consumption. Lakeside fishing communities have lost species that traditionally provided food and supported the local economy. The long-term impacts remain to be seen, but this example provides a valuable lesson for future introductions and transfers of fish species.

Tilapia

Tilapias, consisting of species of the genera *Tilapia, Oreochromis* and *Sarotherodon,* have been widely distributed around the world from their original African home. They are now the mainstay of small-scale aquaculture for many poor farmers in the developing world, as well as for enterprises in the developed world. They are most widely cultured in Asia, particularly China, the Philippines and Thailand.

Dubbed the "aquatic chicken", tilapias possess many positive attributes that suit them for a wide range of aquaculture systems: excellent growth rates on a low-protein diet; tolerance of a wide range of environmental conditions; high resistance to diseases and parasitic infections; ready breeding in captivity and ease of handling; and wide acceptance as food fish. Because tilapias are so widely farmed in the developing world, the Philippines-based ICLARM, the CGIAR centre devoted to fisheries, has established the Genetic Improvement of Farmed Tilapia (GIFT) programme. Its aim is to increase food production and income by and for small-scale producers.

The GIFT programme has collected strains of tilapia and evaluated their culture and growth in different environments. Scientists have discovered, for example, that tilapia breeds in Asia are deteriorating as a result of generations of inbreeding. Future breeding efforts must draw on a wider genetic base, incorporating genetic material from Africa. This underscores the importance of future conservation and utilisation of Africa's native tilapia breeds.

Responsible Fishing

In May 1992, the International Conference on Responsible Fishing at Cancún, Mexico, called upon FAO to draft, in consultation with other international organisations, an International Code of Conduct for Responsible Fishing. The concept of "responsible fishing" embraces sustainable utilisation of fisheries resources in harmony with the environment, and the use of capture and aquaculture practices that do not harm ecosystems, resources or food quality.

FAO supports comprehensive programmes on fisheries management, focusing on both coastal zones and high seas. It is also committed to international efforts to introduce ecologically safe fishery technologies. FAO provides technical assistance aimed at environmentally sound aquaculture practices, as well as incorporating aquaculture in rural development planning.

To conserve aquatic biodiversity, FAO emphasizes the sustainable use of aquatic resources. Activities include genetic selection programmes in aquaculture; the elaboration of codes of practice for the introduction and transfer of aquatic organisms and on access to genetic resources and biotechnology; and maintenance of a world database on introductions and transfers, as well as a database on species, strain and race identification.

Biodiversity Conservation on Forests

About 30 percent of the world's ice-free land surface is forest or woodland. Forested areas of the world today comprise between 3 000 million and 3 500 million ha — an area equal to the size of North and South America. According to recent estimates, temperate forests cover approximately 1 430 million ha in the industrialised countries and another 210 million hectares in non-tropical developing countries. Tropical forests, both moist and dry, cover an estimated 1 760 million ha.

Forests supply food, fodder, medicine and timber, poles and fuelwood as well as raw materials for industry. The income earned from trees and forests is of vital importance to both rural populations and national incomes. Forests are home for an estimated 300 million people — shifting cultivators and hunter-gatherers —

around the world. In the past, the slash-and-burn agriculture practised by forest-dwelling people was sustainable, but population pressures are reducing the land available for shifting cultivation; shorter fallow periods and overuse are turning traditionally sustainable methods into destructive ones.

Rural people living in and around forest areas depend on a large variety of forest products for subsistence. Forest foods form a major part of the diet of some population groups in rural areas in developing countries. They include leaves, seeds and nuts, fruit, roots and tubers, sap and gums, fungi and animals. Forest foods often increase in importance during the hungry season, which reaches its peak just before crops are harvested, and when crops fail.

Woody species provide three-quarters or more of the population in developing countries with their primary energy source. In developing countries, eight times more wood is used for fuel than is logged for industrial purposes. In many areas, fuelwood is being harvested faster than it is being replenished. By the year 2000, nearly 3 000 million people could face fuelwood shortages.

Forests provide vital ecological functions. Their absorption of carbon dioxide and release of oxygen through photosynthesis help control the level of greenhouse gases and provide an atmosphere essential to support life. Forest vegetation helps recycle nutrients. Forest cover also reduces soil erosion by slowing the runoff of water, reducing the hazard of floods and the silting of reservoirs and waterways.

Forests, woodlands and other wilderness areas are increasingly valued as sites of natural and cultural heritage, as well as for education and recreation.

Ecotourism is the third most important source of income in Rwanda, for instance, largely because it is home to the mountain gorilla. The world's forests are declining at unprecedented rates. Major threats are deforestation and atmospheric pollution. Another threat is the narrowing of the genetic base of tree species as a result of commercial forestry operations.

Whereas reforestation of temperate forest lands now exceeds removal of trees, the loss of tropical forests gives cause for concern. The tropical forests were destroyed at an annual rate of 15.4 million ha between 1980 and 1990 according to a recent FAO survey. In terms of area, the greatest losses were in Latin America and the Caribbean (an average of 7.4 million ha per year) followed by Africa (4.1 million ha per year) and Asia and the Pacific (3.9 ha per year).

The causes of deforestation vary from region to region. The most important include: conversion of forest land to agricultural use; excessive use of fuelwood and charcoal; shifting cultivation where fallow periods are too short; unsustainable logging; expansion of urban and industrial areas; and overgrazing and fodder collection. Poverty is the underlying cause of many of these environmentally degrading activities.

Despite a net increase in the forested area in Europe, pollution and forest fires have caused a severe decline in biodiversity and forest vigour. Forests in Germany and the former Czechoslovakia have been particularly affected. Less obvious, but equally alarming, is the decline in genetic diversity within forest species in both Europe and North America. This genetic erosion results mainly from deforestation, compounded for a few economically important species by intensive breeding for commercial forestry. FAO estimates that about 400 tree species are endangered in whole or in significant parts of their gene pools.

When forests decline or are removed, much more than trees is lost. Forests harbour many animals and plants that depend on their environment for survival. Many of these species, their potential value to society and their ecological importance have yet to be discovered. Untapped treasures include possible crops, pharmaceuticals, timbers, fibres, pulp, soil-restoring vegetation, petroleum substitutes and countless other products and amenities. The bark of the rare western yew tree *Taxus brevifolia,* which is now found only in the old-growth coniferous forest of the northwestern United States, was recently found to be the source of taxol, one of the most potent anticancer substances ever found. If forest felling continues at the present rates, new sources of scientific information are likely to be lost and inestimable biological wealth destroyed.

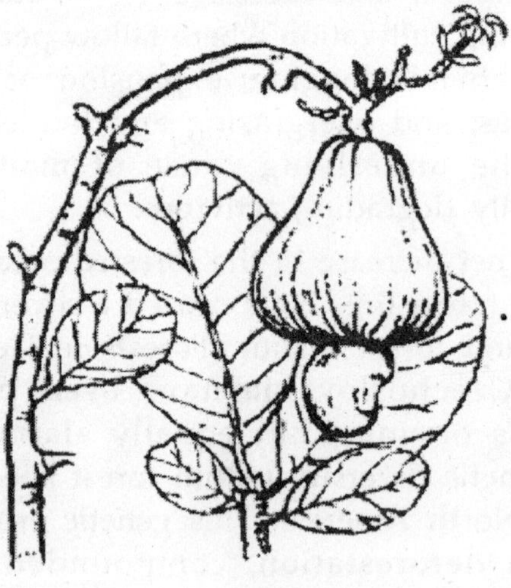

The cashew nut, a nutritious forest Product

Even where conservation measures have been taken, they may not halt the decline in biodiversity and therefore the overall genetic resources of the forest ecosystem. At present less than 5 percent of the earth's land surface is allocated for conservation as national parks, scientific stations or other types of legally protected land. Conservation areas have been set aside for many reasons, but rarely with reference to the location of valuable gene pools. Frequently they are too small to maintain viable populations of the threatened species and varieties they do contain. At the same time, experience shows that policies to control and protect such reserves will not succeed without the active support of local people and complementary programmes aimed at meeting their everyday needs.

Sustainable Development

Properly managed, forest ecosystems can provide goods and services while, at the same time, perpetuating the genetic resources contained in them. Progress is being made towards new styles of management. The sustainable harvesting of non-wood forest products can improve food security and nutrition, while increasing income and job opportunities. Agroforestry — a farming system that combines trees, crops and livestock — enables farmers, even the poorest, to diversify agricultural production and reclaim degraded land. The degradation of forests can also be reduced by harvesting practices that enable logging to take place while promoting and conserving forest regeneration. The sustained utilisation of forests, coupled with the maintenance of a network of areas dedicated to the protection of ecosystems and their functions, provides the only solution for lasting genetic conservation.

Conservation of Forest Genetic Resources

The FAO Panel of Experts on Forest Gene Resources, established 25 years ago, guides the Organisation's actions to conserve forest genetic resources. FAO's Forestry Department collaborates with national or regional institutes that are or wish to become involved in these activities. Its field projects offer technical advice and assistance to governments in planning and carrying out conservation projects, as well as the integration of genetic resource conservation in land-use and forestry planning.

Specific activities include assistance in the exploration, collection and evaluation of forest genetic resources, planning and developing seed centres, and establishing and managing the conservation, both *in situ* and *ex situ*, of priority species. The FAO Forestry Department also publishes and disseminates a wide range of educational and training materials on the use and conservation of forest genetic resources.

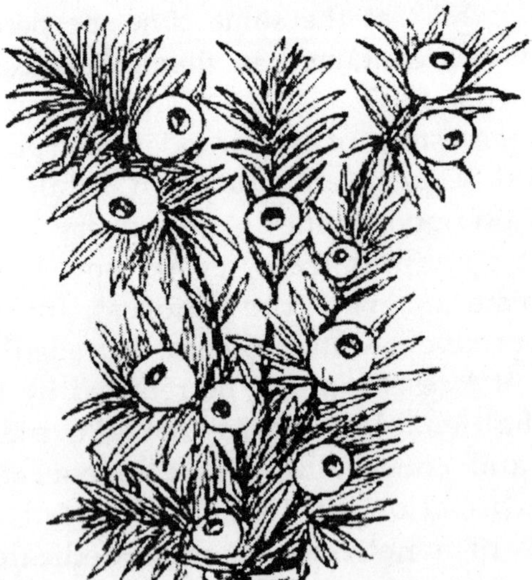

Western yew tree, the source of taxol, a potent anticancer chemical

FAO's Global System for the Conservation and Utilisation of Plant Genetic Resources includes forest tree species. Within the framework of FAO's International Undertaking on Plant Genetic Resources, FAO's Forestry Department is the focal point for activities related to *in situ* conservation of plant genetic resources.

10

Pesticides in Agriculture

In agricultural industry, pesticides are an important component of economic and effective pest control and their continued use is essential. With the benefits of increased knowledge and experience, it is apparent that these products must no longer be used as they were in the past. All farm chemicals must be utilised strategically in the farming system and only be applied with care by competent operators. We all benefit from farm chemicals and therefore we all have a responsibility to ensure that these benefits are maximised, while any adverse effects are minimised.

Pesticides (or farm chemicals, agro chemicals or agvet chemicals) are those substances which are used to control, destroy, repel or attract pests in order to minimise their detrimental effects. Pests are those organisms like weeds, insects, bacteria, fungi, viruses and animals which adversely affect our way of life. Pests can reduce the quality and quantity of food produced by lowering production and destroying stored produce; they can harm our animals (like fleas, worms and diseases); they compete with humans for food and affect the health, welfare and way of life of people; they can destroy buildings (termites) and are a major cause of land degradation (noxious weeds, rabbits, feral pigs, etc). Pests are also a major nuisance around our homes (prickles in the lawn, flies, etc). Pest activity greatly increases the costs of farming.

Pesticides therefore are used in many situations such as livestock farming, cropping, horticulture, forestry, home gardening, homes, hospitals, kitchens, roadsides, recreational and industrial areas. They are a vital facet of farming and our everyday life. Pesticides may be derived from inorganic sources (copper, sulphur), natural organic sources (plants) or be organic compounds synthesised in a laboratory. Many of these synthesised products mimic the activity of natural organic compounds.

While the first recorded use of chemicals to control pests dates back to 2500 BC, it is really only in the last 50 years that chemical control has been widely used. Many of the earliest pesticides were either inorganic products or derived from plants, for example burning sulphur to control insects and mites. Other early insecticides included hellebore to control body lice, nicotine to control aphids, and pyrithrin to control a wide variety of insects. Lead arsenate was first used in 1892 as an orchard spray while about the same time it was accidentally discovered that a mixture of lime and copper sulphate (Bordeaux mixture) controlled downy mildew, a serious fungal disease of grapes. It is still one of the most widely used fungicides.

Many of these early chemicals had disadvantages. They were often highly toxic, were very persistent, posing a threat to the environment, or they damaged the crops they were meant to protect. The modern era of chemical pest control commenced during World War II. For example, the much maligned DDT played a major role in the health and welfare of soldiers who used it to control body lice and mosquitoes which transmitted major illnesses.

Further developments of insecticides and herbicides followed. With their relatively low cost, ease of use and effectiveness, they became the primary means of pest control. Protection of crops, produce, animals and humans over extended periods became possible with corresponding increases in food production and improved standards of living. Modern pesticides are sophisticated compounds which are very carefully researched to ensure they are effective against target organisms, are safe to the environment and can be used without undue hazards to the operators or consumers. Many of these have been developed to target specific biochemical reactions within the target organism, e.g. an enzyme necessary for photosynthesis within a plant or a hormone required for normal development in an insect. Modern chemicals are much safer, more specific and friendlier to the environment than the older products they have replaced.

Advantages of Using Pesticides

Pests are an ecological problem and therefore our control strategies must be ecologically sound. Basically, there are two approaches to pest control:

— prevention

— cure or removal of the cause.

Modern agriculture is a combination of both and human intervention is necessary, whether it be pulling out weeds by hand, use of pesticides or genetic engineering. Control methods evolve over time as knowledge and techniques improve. This includes the development of chemical means of control which become very important because of a number of advantages. For example:

Cost effectiveness. Farm chemicals are an economical way of controlling pests. They require low labour input and allow large areas to be treated quickly and efficiently. It has been conservatively estimated that for every dollar a farmer spends on farm chemicals he receives $4 return. Production per labour unit has increased while production costs and energy inputs are lower.

Timeliness and flexibility. A suitable farm chemical is available for most pest problems with variations in activity, selectivity and persistence. The best product can be chosen for the situation. This allows more flexibility in management options and better timeliness of pest control.

Quality, quantity and price of produce. Farm chemicals ensure a plentiful supply and variety of high quality, wholesome food at a reasonable price. Modern society demands nutritious food free from harmful organisms and blemishes. Ornamental horticulture also requires unblemished and pest-free plants and flowers. This would be very difficult without farm chemicals.

Prevention of problems. Farm chemicals are frequently used to prevent pest problems from occurring, e.g. preventing weeds in gardens and lawns; treatment of export and import produce to prevent the spread of pests; treatment of stored products to prevent pest attack and destruction during storage.

Protection of pets and humans. Without farm chemicals the treatment of spiders, cockroaches, etc in houses; fleas on pets, etc, would be most difficult.

Protection of the environment. If no farm chemicals were available to control environmental pests like noxious weeds, feral animals, etc, our environment would suffer very badly. Using herbicides to control crop weeds reduces the need for cultivation, thus reducing land degradation.

Effects of Pesticides

Despite their many advantages, there are some potential hazards or risks when using farm chemicals. These risks may be associated with all chemicals whether they be industrial chemicals, pesticides, household products or even natural chemicals found in the environment. Undesirable side effects of farm chemical use usually stem from a lack of understanding of the impact of the chemical on the environment, compounded by indiscriminate and overuse of the product. These side effects do not always occur when farm chemicals are used and damage does not necessarily result. Some of these effects may be:

1. *Reduction of beneficial species*. Non-target organisms, including predators and parasites of pests, can also be affected by chemical application. The reduction of these beneficial organisms can result in changes in the natural biological balances. Losses of honeybees and other pollinating insects can also be a problem.
2. Drift of sprays and vapour during application can cause severe damage and residue problems in crops, livestock, waterways and the general environment. Care in the methods of application and the weather conditions under which it is carried out can reduce drift. Environmental pollution from careless application and runoff can result in wildlife and fish losses. This should be a concern for all of us.
3. Residues in food for humans and feed for livestock can be a consequence of direct application of a chemical to the food source, by the presence of pollutants in the environment or by transfer and biomagnification of the chemical along a food

chain. Not all residues are undesirable although good agricultural practice must be observed to prevent unnecessary and excessive levels of residues.

4. Ground water contamination by leached chemicals can occur in high use areas if persistent products are used.
5. Resistance to the pesticide used can develop in target pests due to overuse and incorrect use of the chemical.
6. Poisoning hazards and other health effects to operators can occur through excessive exposure if safe handling procedures are not followed and protective clothing not worn. Poisoning risks depend on dose, toxicity, duration of exposure and sensitivity.
7. Other possible health effects due to indiscriminate use of farm chemicals also concern many people in the community.

Problems result from misuse, abuse and overuse. Farm chemicals can be used safely and effectively without these undesirable effects although there is always a risk associated with any activity. This risk is relative, as evidenced by smoking or driving a car or a tractor. Despite the relative risks being low, all users of farm chemicals, whether on a large scale or in the home, have a responsibility to use them correctly.

Farm Chemicals

Farm chemicals are important because they enable us to produce sufficient high quality, wholesome food for a growing population. They also allow efficient and economic pest control and often there is no alternative control option. People who enjoy a healthy lifestyle because they have unrestricted access to an abundance of good, inexpensive food easily overlook the fact that historically starvation has been a major health hazard. Civilisation has been combating weeds, insects, diseases and other pests throughout history and there are many examples of how these pests have had a major impact on humans.

One of the worst examples is the Black Plaque of Europe in the fourteenth century when millions died from a bacterial disease spread by fleas from rats. Another example is the infamous Irish potato famine of the nineteenth century in which millions died and many more were forced to emigrate. A fungus also destroyed the entire German potato crop in the early twentieth century resulting in 700,000 deaths from starvation.

In many countries similar situations continue. Economically advanced countries have plenty of good wholesome food due to a scientific approach to agriculture which includes the use of pesticides. Without the use of farm chemicals, the production and quality of food would be severely jeopardised with estimates that food supplies would immediately fall to 30 to 40% due to the ravages of pests. While there are mountains of food in Europe and the US, this represents only 45 days food supply for the world.

A recent United Nations report stated that population growth is a major problem facing our planet. In 1900 there were 1.6 billion people on the planet. In 1992 this had risen to 5.25 billion and by the year 2050 it will reach 10 billion. World population is increasing by 97 million per year. By the year 2000, 50% of world population will live in cities and 75% will live in developing countries. This explosive increase in world population is mostly in developing countries and this is where the need for food is greatest. FAO estimates that 500 million are already undernourished. There is a limit to new areas to cultivate, therefore we must increase agricultural production from the areas available.

Benefits of Farm Chemicals

To maximise the advantages and benefits of farm chemicals while minimising potential problems. This can be achieved in three ways:

i) by ensuring there are adequate safeguards over the manufacture, sale and use of these products;

(ii) by ensuring farm chemicals are only used with an integrated pest management (IPM) approach;

(iii) by ensuring farm chemicals are used correctly, safely and accurately by well trained and competent applicators.

Each of these will be considered in turn.

Safeguards

There are very stringent controls and regulations governing all aspects of farm chemicals, including:

— development and testing

— registration

— transport, storage and sale

— use

— residues in food

— disposal of waste

— environmental contamination

It is important to understand the development and testing procedures to which modern pesticides are subjected, and the controls placed on their use. A company cannot simply develop a chemical, call it a pesticide and immediately sell it. A long and costly process is involved. Out of every 10,000 compounds developed by chemists, only about one completes the development phase and satisfies the many controls that have to be passed before it can be offered for sale. To gain registration all public health, occupational health, environmental and agricultural concerns must be satisfied.

All tests are conducted to international standards. All agricultural and veterinary chemicals are subject to ongoing review at all times thus ensuring that new scientific information can be taken into account as soon as it becomes available. Should this information indicate a problem with a chemical product then its use may be severely restricted or its registration withdrawn thus making it illegal to sell or use the product. As well as stringent controls over the development and sale of pesticides, there is also strict regulation of their use. This regulatory system aims to protect human and environmental safety, ensure product effectiveness thereby avoiding unnecessary chemical use, and to safeguard Australia's export trade by ensuring our produce is free of pests and also free of unwanted chemical residues.

With these levels of control, we can be confident that commercial farm chemicals are effective and can be used with minimum adverse effects, and our food is free of undesirable residues provided the chemicals are applied correctly and the directions on the label are adhered to. In fact, pesticides are tested to the same levels as human medicines. This is an area the organic farming movement needs to urgently address. Simply substituting one product for another does not constitute organic farming and naturally occurring chemicals are not necessarily better than synthesised ones. They are all chemicals.

Competency

Competency is a combination of knowledge, skills and attitude. Farm chemical users need to know what they are doing and have a sound knowledge of the products they are using. They then need to have the skills to ensure this knowledge is used to best advantage, while the correct caring attitude is necessary to safeguard themselves, their families and the community. Such competency can best be achieved through training and accreditation.

Accurate use of pesticides

It is essential that pesticides are not used as the sole means of pest control. Misuse, overuse and abuse leads to many of the problems previously discussed. Pesticides must be used as part of a planned systematic pest management program utilising as many control techniques as applicable. This is called Integrated Pest Management (IPM). Controls such as biological, physical, cultural, genetic, environmental and mechanical techniques are just as important as pesticides. The IPM philosophy and approach is the only way to economically and ecologically sustainable farming. When pesticides are used, it is vital they are used correctly. Some of the considerations involved are:

— selection of product
— compatibility with other controls
— safe work practices
— following label directions
— timing of application

— accurate application to the target

— selection, adjustment and calibration of equipment

— application under favourable weather conditions

— keeping good records

By correct, safe and accurate use of these products, better pest control will result, adverse effects will be minimised, and farming will be more profitable.

Integrated Pest Management (IPM)

In agriculture, Integrated Pest Management (IPM) is a pest control strategy that uses an array of complementary methods: natural predators and parasites, pest-resistant varieties, cultural practices, biological controls, various physical techniques, and pesticides as a last resort. It is an ecological approach that can significantly reduce or eliminate the use of pesticides.

An IPM regime can be quite simple, or sophisticated enough to be a farming system in its own right. The main focus is usually insect pests, but IPM encompasses diseases, weeds, and any other naturally occurring biological crop threat.

An IPM system is designed around six basic components:

Acceptable pest levels: The emphasis is on *control*, not *eradication*. IPM holds that wiping out an entire pest population is often impossible, and the attempt can be more costly, environmentally unsafe, and all-round counterproductive than it is worth. Better to decide on what constitutes acceptable pest levels, and apply controls if those levels are reached.

Preventive cultural practices: Selecting varieties best for local growing conditions, and maintaining healthy crops, is the first line of defense.

Monitoring: Regular observation is the cornerstone of IPM. Visual inspection, insect traps, and other measurement methods are used to monitor pest levels. Record-keeping is essential, as is a thorough knowledge of the behavior and reproductive cycles of target pests.

Mechanical controls: Should a pest reach an unacceptable level, mechanical methods are the first options to consider. They include simple hand-picking, erecting insect barriers, using traps, vacuuming, and tillage to disrupt breeding.

Biological controls: Natural biological processes and materials can provide control, with minimal environmental impact, and often at low cost. The main focus here is on promoting beneficial insects that eat target pests.

Chemical controls: Considered as an IPM last resort, synthetic pesticides may be used when other controls fail or are deemed unlikely to prove effective. Biological insecticides, derived from plants or naturally occurring microorganisms (eg: Bt), also fit in this category.

IPM is applicable to all types of agriculture. Reliance on knowledge, experience, observation, and integration of multiple techniques makes IPM a perfect fit for organic farming (the synthetic chemical option is simply not considered). For large-scale, chemical-based farms, IPM can reduce human and environmental exposure to hazardous chemicals, and potentially lower overall costs.

Biological Pest Control

Biological pest control is a rapidly expanding field of agriculture, where natural agents, primarily parasitoids and predators are used to control a pest organism that has been causing economic harm to human interests. These methods can be as alternatives or supplements to conventional pest control methods such as insecticides. Human disturbance of ecology ecological systems, particularly the development of monoculture and the movement of species tends to stimulate pest problems by reducing biodiversity which usually has natural controls that keep pest populations in line. Biological pest control does not seek to erradicate pests, but to keep them to economically minimal factors by reducing populations to acceptable levels.

Some organisms have been intesively studied and are in modern commercial use. Many other organisms have potential and are candidates for use in this field. This category includes both current and potential biologicial control agents. The misuse of pesticides is harmful to human health and is damaging to the environment. Millions of people are poisoned by pesticides every year, and pest problems are often made worse when the balance between beneficial and harmful insects is disturbed by applying toxic chemicals.

In the last two decades Asian Governments, NGOs and international agencies have been working together to organise training which helps farmers to learn about the ecology of their fields and, as a result, enables them to make and implement decisions which are safe, productive and sustainable. This ecological approach to plant protection is called Integrated Pest Management, or IPM. Not only does it involve minimising the use of pesticides, it also involves a wide range of other practices aimed at growing a healthy crop.

The training approach which has been used to help rural people learn about IPM is called the Farmers Field School (FFS). This entails weekly meetings by a group of farmers. Instead of listening to lectures or watching demonstrations, these farmers observe, record and discuss what is happening in their own fields from the time of planting to the time of harvest. This discovery-learning process generates a deep understanding of ecological concepts and their practical application.

Since 1990 more than two million farmers have graduated from FFS. In recent years, IPM farmers have started organising themselves in order to carry out field experiments, train other farmers, and interact more effectively with government agencies. These developments have given rise to a new term, Community IPM.

11

Chemicals and Food Safety

The foods we consume at times contain both natural and synthetic chemical compounds that are intrinsically toxic. Also present in foods, however, may be detoxifying compounds which, together with detoxification processes in our bodies, limit the damage inflicted on our metabolism. We are usually aware of the more acute responses which arise from allergens or microbial toxins, but are less aware of compounds which cause chronic diseases over time.

Food Hazards

The risk to health of various food hazards has been examined objectively, based on the severity, incidence and onset of biological symptoms. The six principal categories of food hazard, ranging from greatest to least, are follows:

1. Microbial contaminants
2. Nutritional imbalance (excess and deficiency)
3. Environmental contaminants
4. Natural toxicants
5. Pesticide residues
6. Food additives

The ranking of these hazards however, is not linear, that of environmental contaminants and natural toxicants being about 1/1000th that of nutrient imbalance and pesticide residues and food additives being about 1/100th that of natural toxicants. This scientific perception is almost the reverse of the public's perception of risk where pesticides and food additives are seen as the greatest dangers.

Natural Toxicants

Chemical compounds found naturally in foods can be toxic in a variety of ways. They can affect the nervous system, cause intestinal and liver damage, impair hormonal function or cause mutagenesis and carcinogenesis. The glycoalkaloids -chaconine and -solanine are examples of toxic compounds found in potatoes and other members of the Solanaceae which interfere with normal function of the central nervous system and irritate the gastrointestinal tract. Acute illness and even death has followed the consumption of damaged, green or blighted potatoes. The active ingredients affect cholinesterase activity and symptoms of illness include laboured breathing, weakness, paralysis and lack of consciousness. The effects on the intestines lead to inflammation of the intestinal mucosa, ulceration and pain.

The brassica vegetables such as cauliflower, cabbage and Brussel sprout contain a complex class of sulphur containing glycosides known as the glucosinolates. These compounds can break down to form a range of products which are toxic to animals and man. They can affect growth and performance in livestock by causing liver damage, impairing normal thyroid function and interfering with other metabolic processes. Less attention has been paid to their effects in man. Some products are known to be hepatotoxic or have slight mutagenic activity while two are known to interfere with either the formation and secretion of thyroxine or with the binding of that hormone. Others, however, are thought to possess tumour suppressant activity.

Saponins, which are found amongst widely eaten legumes, act as dispersants and increase the permeability of the intestine to a wide range of toxicants such as tannins, enzyme inhibitors, phytate and oestrogenic isoflavones. This has been observed in experimental animals and clinical investigations need to be carried out in man to determine if there is synergism between saponin consumption and the biological activity of absorbed toxins.

Dietary Mutagens and Carcinogens

In relation to mammalian carcinogenesis, there are many natural compounds found in foods which have the potential to be at least mutagenic and thus pre-carcinogenic. These compounds range from actual chemicals found in some plants, such as hydrazines in mushrooms, to those produced by some mycotoxins, associated with plant products, such as the aflatoxin in peanuts. An extension of this is those toxins produced on cooking, notably the pyrolysis products of cooked meat.

Apart from the hazards of natural compounds, the residues of some chemicals used in agricultural production are of some concern if misused, as the active ingredients of some pesticides, for instance, are carcinogenic. At the recommended levels of application, the maximum residue levels (MRL) of these chemicals found in foods are such that thorough testing by industry and health authorities suggest that they are safe. There remains great public concern however, almost hysterical fear in some quarters, that foods produced in this way are harmful particularly in regard to their cancer causing effects. Unlike some carcinogenic testing in the past where massive dosing has been used, these procedures expose experimental mice to small quantities of material over long periods.

Detecting Genotoxicity

Free superoxide radicals can arise from the detoxification of natural and synthetic toxins and to a degree from normal metabolism. Radiation and oxidants from pollutants can also be a source of these compounds which, after some modification, are able to enter the cell nucleus and damage DNA material. The body uses deoxygenating enzymes and antioxidants to limit the input of these potentially dangerous compounds and thus diets rich in antioxidants are recommended for good nutrition.

The DNA of the mammalian chromosome can be damaged in a number of ways. Single or double strand breaks can affect the DNA molecule of the chromosome and, in addition, bases can be damaged, deleted or modified by the covalent binding of mutagenic compounds (adducts). In other circumstances, strand breakage and reunion can involve exchange of material between individual chromosomes (recombination). This sort of event can lead to loss of tumour suppressor genes or activation of oncogenes which in turn may lead to cancer initiation or progression. In man there is always present a spontaneous mutation rate, but when man or animals are challenged by certain toxicants these rates increase. It is now possible to measure this increment.

In one assay, mutations can be detected by the presence of micronuclei which arise from small portions of fractured nuclei. In others, measurements in lymphocytes are made by tissue culture techniques of point mutations at x-linked loci or, alternatively, of mutations, deletions and in some cases recombinations at autosomal loci. Mutant lymphocytes are isolated by tissue culture and analysed by molecular biology techniques.

Natural Antimutagens and Anticarcinogens

Offsetting the effects of mutagenic agents are the antimutagenic/anticarcinogenic action of a diverse range of chemicals found in a variety of foods. These include the antioxidant and free radical scavengers such as tocopherol found in cereals, and vitamin A and C found in the orange and green vegetables and citrus, but many others are important also. Fibre is seen as important in that it can bind carcinogens, both internally generated (e.g. bile salts) or others consumed in the diet.

Genotoxicity and Public Health

In the past, food safety testing to a large degree has examined acute responses to foods or specific chemical compounds. The battery of tests now available to assess the genotoxicity of food components will enable us to assess the dangers which arise from the long-term consumption of mutagenic compounds. Despite the public concern over chemicals used by man, there still exists doubt about the properties of many natural compounds. In both cases, genotoxic testing should, firstly, recognise unsafe compounds and, secondly, provide scientific evidence on the protective effects of antimutagenic agents found in natural foods.

Examining Food Safety

Over the past two years one issue came top of the list month after month. The overwhelming impression gained by those responding to the queries and the complaints was that consumers somehow feel that the food supply is out of their control and it should not be. In the past we knew exactly what we were eating. We grew some of it in the back yard, we bought some from the greengrocer, some from the butcher and the butter, sugar and flour from the grocer. Today we are confronted by a vast array of brightly packaged designer foods offering us convenience, fun and happy families forever.

Consumers are not sure what is in the food they are offered. In the media they learn that their pork may be genetically modified, their spices irradiated, and their cheese contaminated with listeria. They believe that there are more food additives in the food than is good for them and that if overseas countries are rejecting our meat, then there must be something seriously wrong. Consumers are confused by conflicting nutrition messages and feel helpless in combating the pressures of children subjected to high fat/high sugar food advertising.

The major types of chemical residues which may be found in our foods today.

Hormonal Growth Promotants

The current lack of knowledge of the long-term effects on humans, although at this stage there is no conclusive evidence of risk when used correctly. Doubts have also been raised over proposals for the use of Bovine Somatotropine (BST), a genetically engineered growth hormone claimed to boost milk production by 10 to 25 per cent.

Antibiotics

The use of antibiotics to control infection and promote growth is widespread in livestock production. It would be fair to say that consumers have few problems in principle with using veterinary chemicals for prevention or treatment of disease, assuming, of course, that the correct drugs and dosages are being used and the withholding periods observed. There is, however, substantial consumer opposition to the use of these drugs as growth promotants.

Pesticides

Consumer organisations still have some concerns over the reluctance to establish specific pesticide residue limits in the final products, but regard the standard as the first step in providing consumers with the knowledge to make an informed choice about organic produce at the point of sale. The organic fruit and vegetable growers have demonstrated that there is a market for organically grown produce. The challenge is now ready for producers of other commodities to move into this market.

Consumers have demonstrated that they are prepared to pay for safety and for quality in food. They need to be reassured that they are getting exactly that. Consumers must rely on the integrity of the voluntary labelling systems in identifying organic foods. While self-regulation by the industry itself is in force, all of those in the industry must ensure that there is constant monitoring of products in the market place, the enforcement of rigid standards and the threat of effective sanctions for those who do not comply. It will only take one retailer or one processor to abuse the system and all the trust and goodwill will disappear, together with the organic food market.

12

Energy Function in Agriculture Sector

The concept of Sustainable Agriculture and Rural Development (SARD) aims to foster sustainable development in the agriculture, fisheries and forestry sectors that conserves land, water, plant and animal genetic resources, and is environmentally non-degrading, technically appropriate, economically viable and socially acceptable. Attaining food security requires policies that ensure social, cultural, political, and economic stability. Combining the economic, social and environmental functions of agriculture can help to achieve these goals.

Agriculture can deliver a wide range of non-food goods and services. This can include its use as a viable, sustainable source of energy, as set out in the FAO perspective on the "Multifunctional Character of Agriculture and Land (MFCAL)". The perspective recognises that agricultural activity and related land-use contribute directly to other, non-agricultural functions comprising social, environmental, economic and cultural goods and services. These can result in significant benefits or costs. Evidence suggests that, in addition to food security, agriculture makes a major contribution to achieving sustainability in rural development, energy and the environment at local, national and global levels. The effective operation of the market stimulates the identification and enhancement of these multiple functions and the emergence of new techniques and technologies.

Plant and crop-based resources are used as raw materials for a wide variety of industrial products, ranging from wood used in paper and board manufacture, starches used in adhesives and vegetable oils used in paints and resins. Agriculture provides a key economic sector for the supply of carbo-hydrates, lignins and plant oils, and this supply is renewable over a short time frame compared with the supply of hydrocarbons from fossil fuels. In several industrialised countries there are steps being taken to enhance the use of plant and crop based renewable resources through technological developments.

Energy is needed for activities such as mechanisation, water pumping, irrigation, fertiliser production, transport and food processing and storage. The role of agriculture as a major energy supply sector is rarely recognised or put into practice. Awareness of the potential for bioenergy as an economic driver for rural development, together with growing attention to global climate change have highlighted this new approach to the energy function of agriculture.

The traditional rural energy sector in developing countries is already largely based on biomass from agricultural and forestry sources. Woodfuel, especially charcoal, is already very much a traded commodity, and farmers can earn extra income from sale of woodfuels.

Energy Supply

The production and use of biomass as energy sources are

linked to many issues, including agriculture and food security, land use and rural development, sustainable forest management and biodiversity conservation, and mitigation of climate change. Bioenergy must also be seen in relation to poverty, population development and health. The fact that women and children in many rural areas spend a good portion of their working day in search for fuelwood, reflects the need to look at bioenergy in the context of gender roles and survival strategies for the poorest of the poor.

The imbalances between household economy and the environment needs to be resolved, together with conflicts between conservation and consumption of biomass and between the present and future needs of societies. It is also important to note that ways of producing and distributing conventional energy are changing as a result of new approaches such as privatisation, decentralisation, trade liberalisation and globalisation.

Access to adequate and affordable energy is one of the prerequisites for equitable socio-economic development, and use of biofuels can contribute towards a more gender balanced rural employment and income, and can strengthen rural livelihood systems to attain better levels of food security. New and/or improved technologies for bioenergy utilisation as an industrial energy source at competitive market prices can contribute towards an improvement of rural livelihood systems.

Set against this approach is the potential threat to forests and trees outside forests if fuelwood is used in an indiscriminately and unsustainable way, which can result in forest degradation or deforestation, deterioration of watersheds, loss of soil fertility as well as biodiversity. Possible conflicts with other land use requirements must also be resolved. Nevertheless the analysis indicates that a substitution of fossil fuels through an increased utilisation of biofuels, produced in a sustainable manner, can contribute towards a cleaner environment, reduction of emissions and mitigation of climate change.

Production of Energy Crops

Land availability is often seen as a constraint to the production of energy crops. With many people in developing countries still undernourished, it is a justified concern that there should be sufficient land for food production and that food should be the priority. However, food production is a complex socio-economic, political and cultural issue that goes beyond the earth's carrying capacity to grow food crops. If farmers are given the opportunity through economic incentives, land tenure rights and capital investment, they will be able to produce more food than has been the case so far.

In parallel with the prospects of increased food production, there are large areas of deforested and degraded land that would benefit from the establishment of biomass plantations, with estimates ranging up to over 300 Mha available for reforestation and agro-forestry. Other studies of the potential cropland resources in developing countries have indicated that these countries will be using only 40% of their potential cropland in 2025. The balance between higher yields in good lands and the benefits of bringing back into production degraded lands is an important issue.

Bioenergy programmes, when coupled with agro-forestry and integrated farming, have the potential to improve food production by making both energy crops and income available. Increasing agricultural production of biomass can be achieved by substituting for other agricultural crops that are in surplus, intermixing energy crops with food or forage crops in an agro-forestry approach, and incorporating into land conservation systems such as windbreaks and shelter-belts. There is also potential to increase the use of crop residues provided this is consistent with the levels of organic matter and control of erosion.

Table 1. Examples of dedicated energy crops

Feedstock	*Conversion technology*	*Products*
Sugar cane	Gasification/existing boiler	Electricity
Sugar		
Switchgrass, wood residues	Gasification/co-firing	Electricity
Sorghum, switchgrass, silver maple, cottonwood	Pyrolysis/combustion turbine	Electricity
Charcoal		
Willow	Co-firing/combustion	Electricity
Alfalfa	Co-firing./gasification combined cycle	Electricity
Animal feed		
Pine	Gasification combined cycle Co-generation/alcohol production	Electricity
Ethanol		
Elephant grass, sugarcane, eucalyptus	Combustion/fermentation	Electricity Ethanol

Table 1 lists some examples of dedicated energy crops, the conversion technologies being developed for their exploitation and the energy and food products that can be produced from them.

Mitigation of Climate Change

The threat of climate change is principally an energy-related problem. Current energy systems are based on the combustion of fossil fuels, which account for 75% of the world's primary energy supply. A second major aspect of the energy function of agriculture, therefore, is the contribution that agriculture can make to climate change mitigation by CO_2 substitution. Carbon sequestration through changes in land management practices are one area of interest, especially as changes such as the introduction of zero-tillage reduce energy consumption in arable farming.

Looking specifically at energy supply issues, biomass offers a carbon-neutral source of energy that is renewable on a short time scale, and hence could provide an attractive means of climate change mitigation. The key aspects that should be considered are:

- the technical potential of CO_2 emissions substitution by the use of bioenergy;
- the technologies that are needed to exploit bioenergy, including their availability, effectiveness and environmental impacts;
- the economics of their exploitation, and the energy markets that can be served by bioenergy technologies;
- the institutional, legal and organisational structures that are required to deliver the assumed benefits.

The Kyoto Protocol is designed to provide binding, quantitative limits on future net emissions of greenhouse gases. Bioenergy could have a major role to play in helping to meet these limits, and mechanisms proposed by the Protocol might make use of agriculture's contribution to climate change mitigation. Whilst the emphasis in international discussions to date has been on forests as a source of carbon sequestration, agriculture also has a major role to play in carbon substitution through the increased use of modern biomass technologies.

Joint Implementation and the Clean Development Mechanism

Joint Implementation (JI) and the Clean Development Mechanism (CDM) are defined under the Kyoto Protocol to the UN Framework Convention on Climate Change as 'flexibility mechanisms' which allow an investor or donor country to fund projects which reduce greenhouse gas emissions (GHG) in a host country. In return the donor country receives 'credits' which contribute to their GHG emissions targets. In JI, a donor country will be an industrialised country with emission targets, whilst the host country will be an economy in transition. In the CDM, a donor country will again be an industrialised country with emission targets, whilst the host country will be a developing country without targets. The credits that will be transferred under JI are called emissions reductions units (ERUs), and those transferred under the CDM are called certified emissions reductions (CERs).

The CDM is of particular relevance to developing countries. A cornerstone of the CDM is that projects implemented through it should assist the host country in achieving 'sustainable development'. CDM projects should also be integrated with national development programmes and be appropriate to the specific conditions of the host country. The essential feature of implementing the CDM will be to balance the aim of contributing to the sustainable development of the host countries with the needs of the donor countries to achieve GHG emissions reductions.

CDM offers great potential for changing the global approach to development, moving this approach away from the Official Development Assistance framework to becoming a much more private sector led framework. If it is designed properly, the CDM can make a decisive contribution to sustainable development and provide a new channel for finance, investment, technology transfer and the promotion of equity. Other features of the CDM include:

- — project activities must be additional to any policy actions that give rise to the same outcomes;
- — it is open to participation by either private or public entities, or combinations of the two;
- — projects must have the express approval of the host government;
- — an executive board will be set up to supervise the operation of the CDM;
- — all CDM projects must be independently certified, possibly making use of operational entities for monitoring and validating emissions reductions;
- — baselines and/or benchmarks will be used to assess the extent of emissions reductions;
- — CERs can be traded between countries;
- — the CDM also has a mandate to use a portion of its proceeds to assist those countries, which are particularly vulnerable to climate change to adapt to those changes.

Projects that are funded via the CDM will be expected to contribute to sustainable development, and renewable energy investments using modern bioenergy technologies are a significant means of achieving this objective. The CDM is due to begin operation from the year 2000, but, at the time of writing, the detailed rules for its operation have yet to be defined. However, several assessments are available of potential projects and their benefits. Some of these studies assessing the value of carbon emissions avoidance by use of renewable energy systems illustrate the economic potential of agriculture's contribution. Comparing the results of these studies with other climate change mitigation measures, such as fuel switching for grid-based electricity generation, and improved energy efficiency in industry enable agriculture's contribution as a carbon-neutral energy resource to be ranked against other emissions control technologies. Nevertheless research in this field is only just starting, and no firm policy conclusions can be drawn, especially as the Kyoto Protocol is not yet in force.

Encouragement of international investment from industrialised countries into developing countries as a means of accelerating renewable energy development may be politically sensitive. Whilst international investment could help local development and lead to greater up-take of certain technologies, it may also expose countries to international competition that could damage indigenous industries and increase dependence on foreign technology. These kinds of impact have particular relevance for renewable energy investments where capital costs are high, and where intellectual property remains in the industrialised world.

It seems clear that technology transfer activities need to be treated carefully and that international organisations have to proceed with caution in stimulating these activities. Whether through the CDM or other flexible mechanisms, foreign investment openings for renewable energy systems, including modern bioenergy, would need to be integrated with robust national industrial and technology development policies. Integrating foreign investment and climate change policy is, therefore, not simply about increasing opportunities for private investors, but is equally concerned with establishing the correct combination of market and regulatory factors that allows investors to earn profits whilst delivering public-sector objectives.

CDM Projects in India

Preparatory studies in evaluating potential CDM projects can help to identify prospects and opportunities. As an example, the World Resources Institute has made an assessment of potential CDM projects in Brazil, China and India, in order to review how they might advance both CO_2 emission reductions and sustainable development. For India, over 20 projects including new technologies for conventional power generation and applications of renewable were examined. All projects broadly advance sustainable development, with benefits ranging from rural electrification and employment opportunities to improvements in productive efficiency.

A set of criteria for evaluating sustainable development benefits against India's development objectives was used and each project was scored against these criteria, which were weighted to reflect the importance of different benefits as determined by polling of researchers and government officials. Of the abatement opportunities reviewed by WRI, there was considerable overlap between projects that offer low-cost CO_2 reductions and projects consistent with India's development priorities. Whilst projects in different sectors should not be compared, renewable energy systems potentially have a large role to play in assisting both CO_2 emissions abatement and sustainable development. Their importance will be dependent on the procedures agreed for the CDM. Furthermore, assessment of CDM opportunities by national governments and private sector investors will require:

— a least cost approach that places bioenergy and other renewable energy options in competition with other mechanisms and policies;

— a comparative financial and technical risk analysis, from which industrialised country investors may prefer certain types of project;

— assessment of the extent to which particular projects contribute to sustainable development objectives;

— the availability of project finance and political support within developing countries to develop specific carbon abatement options.

Role of Bioenergy in Climate Change Mitigation

Bioenergy has played an important part in rural development, but its contribution to an enhanced role in the wider issue of climate change mitigation has not been fully examined in policy formation. Exploitation and commercialisation of biomass through modern technologies offers significant cost-effective opportunities for meeting emissions reduction targets while providing additional economic and social benefits. Bioenergy provides a sustainable use of accumulated carbon and acts as a substitute for the use of fossil fuels.

All forms of biomass utilisation can be considered part of a closed carbon cycle. Biomass utilisation through the substitution of fossil fuels will reduce CO_2 emissions, and a combination of biomass with carbon sink options can provide a viable route to climate change mitigation.

Bioenergy can contribute to sustainable development in developing countries, provided that a number of key issues related to the practical exploitation of this resource are carefully considered. Biomass sources are more spatially dispersed than fossil fuels, and whereas dispersion tends to increase harvest and transport costs, modern biomass technologies have the potential for generating employment and assisting economic growth in rural areas. Current bioenergy sources are mainly forest and agriculture residues, but in the future dedicated energy plantations could provide additional sources, opening up new opportunities for agriculture and forestry in the energy market.

The world-wide potential for energy supply from energy crops, and the specific use of biomass need to be further examined. There are many complex issues in shifting from petro-chemical feed-stocks; for example, the possible conflict with competition for food crops; the economics of bio-energy resource exploitation; and the availability of land for energy crops. In addition, the role of agronomy in matching species to the objectives of growing energy products and to specific local conditions should be examined. For example, there are opportunities for adapting plant species for production of both food and energy crops in drier zones, where 80% of the world's poor live.

Plant matter is associated with three basic crops: cellulosic crops, such as wood and cotton; starch crops (such as corn) and oil crops (such as soybeans). Large-scale displacement of petroleum fuels will rely primarily on low-cost cellulosic feed-stocks, whilst starch crops may play more of a transitional role. Such biomass sources have the potential for renewable energy production. IPCC estimated that biofuel production on 10-15% of the land area currently in agricultural use could substitute for about 784 MtC/year of fossil fuel carbon.

In practice, estimating the potential of biomass to offset CO_2 emissions is complex because of the many variables involved such as crop productivity, energy conversion efficiencies and substitution factors. A simplified approach is to use the product of (*cropland area*)x(*crop yield*) as the main factor in determining carbon offset potential. This calculation should then be scaled by an energy substitution factor to account for conversion efficiencies. Based on this approach, estimates of CO_2 mitigation using bioenergy as a direct substitute for fossil fuels can be made. These are achieved by estimating increases in the land area dedicated to biomass production, the extent of better use of agro-forestry residues, and increased efficiency of biomass conversion processes and integrated food-energy production.

Bioenergy should not be regarded as the only means by which agricultural and energy problems can be solved in rural areas; rather it is an option that can provide some benefits in improving agricultural productivity, giving a renewable source of energy supply and helping to deliver a sustainable environment. Its overall contribution depends on a number of inter-related economic, technical, social and environmental factors. The multifunctional character of agriculture and land should, in fact, receive greater recognition so that effective policies and opportunities can be developed to tackle these rural energy problems.

Potential of Biomass

Biomass offers many potential advantages as an energy supply option; equally there are barriers to be overcome before the full potential can be realised. The balance between advantages and disadvantages is important to assess, and Table 2 sets out these features. It is clear that there remain uncertainties about whether and how bioenergy can be developed and commercialised. Large-scale production would require dedicated energy crop plantations, and the cost-effectiveness in any particular investment situation is likely to depend on site-specific opportunities.

Table 2. Exploiting the Potential of Biomass

Advantages	Disadvantages
— Biomass is a stored fuel, and can provide a demand responsive supply of energy.	— Present-day biomass technology has a poor cost-effectiveness compared with most conventional energy supply options.
— Biomass is a flexible fuel feedstock, which can be converted into convenient secondary forms, namely solid, liquid or gaseous fuels.	— The ownership of modern biomass technology is largely in industrialised countries, and would require technology transfer for its exploitation in developing countries.
— Biomass is a carbon neutral energy source, renewable over a relatively short time-frame	— The long-term effects of biomass exploitation, through dedicated energy crop plantations, on soil quality, fertility and biodiversity may be adverse
— Fuel production, collection and supply are labour intensive, which can be an attractive social benefit in rural areas	—Biomass plantations may come into conflict with other land uses in competition for high quality land
— When developed as part of an integrated approach to wasteland restoration and rural economic development, biomass offers additional environmental, social and economic benefits in rural areas.	—A fuel supply chain and a ready market for energy output, with robust and secure contracts must be in place.

However, the CO_2 mitigation opportunity may tip the balance in favour of biomass investments if they are regarded as being of global importance, especially if GEF funding, or an appropriate value for CERs obtained through the Clean Development Mechanism, can be used to support the investment costs.

Renewable Energy and Development

Renewable energy technologies for developing countries have been the subject of a variety of official assistance schemes. Renewable energy first became an official priority for international aid at the 1981 UN Nairobi Conference on *New and Renewable Sources of Energy*, which called for action in research, planning, investment and dissemination of renewable energy technologies. During the following 10-15 years, there emerged two kinds of initiative: legal or regulatory developments to encourage private investment in renewable energy, and financial assistance to public or private investors from national, bilateral or multilateral sources for capital intensive projects.

However, much of the early international assistance for renewable energy development tended to focus on large-scale projects, and also overlooked the practical needs for promoting technology transfer. New initiatives were designed during the mid-1990s, with the focus changing to an integrated process of encouraging small, local entrepreneurs in developing suitable technologies and then supporting market deployment. More recent experiences of renewable energy programmes suggest that acknowledging local needs and integrating them into technology transfer activities is the best way to ensure that new technologies can be adopted. Policies will also be needed to encourage the private sector, or public-private partnerships, to become involved in improving energy provision in rural areas.

For example, the World Bank has produced a revised set of strategic objectives for its energy policy. These include a larger role for renewable energy systems, creating better links between energy activities and other non-energy sector lending, improving access to modern energy by rural populations to alleviate poverty and projects aimed at reducing the health impacts of traditional fuel use. The Inter-American Development Bank has also included the promotion of new markets for rural energy in its draft energy strategy, and is stimulating renewable energy projects through its Sustainable Markets for Sustainable Energy programme.

In the renewable energy sector, the World Bank launched its Solar Initiative in 1994, which seeks to commercialise renewable energy use in developing countries, and has set up the PV Market Transformation Initiative that aims to accelerate the commercialisation of solar PV energy services. Other initiatives to promote renewable energy by national, bilateral and multilateral agencies have been established in the last few years. Financial and institutional assistance to renewable energy development now tends to address both private-sector needs and improving the regulatory and legislative structure of the energy sector.

Energy-Agriculture Nexus

The energy-agriculture nexus is a coherent system, and that energy, biomass and carbon flows can be identified and traced throughout the system. Biomass is an energy source that is renewable over a short time frame, especially if managed annual crops or perennial and short-rotation woody species are used. Sustained economic growth depends on having a secure supply of energy inputs, and biomass can provide an important feedstock for energy systems in developing countries.

The institutional requirements and the policy frameworks that are needed to build on the nexus between energy and agriculture for developing countries are key factors. For this nexus to be fully exploited, the potential of agriculture as a source of renewable energy for rural development and CO_2 substitution needs to feature more strongly in the policy agendas of national, multilateral and bilateral organisations. The cost-effectiveness of using biomass technologies for energy supply, and the associated social and environmental benefits also should form part of this framework.

Bioenergy can and should play a significant role in improving agricultural productivity, whilst at the same time there is a major opportunity for biomass to contribute both to energy supply and global environmental sustainability. The linkages between the energy, biomass and carbon flow elements should form a main focus of rural development and rural energy policy formulation. Developing the potential for bioenergy has also to be seen in relation to poverty alleviation, rural community development and health. Recognition of the importance of these linkages is now emerging. Technical developments of plant/crop based renewable resources for modern energy conversion systems have been undertaken in many countries over the last 10-20 years, and progress has been considerable. Furthermore, the potential for energy supply from agriculture has gained a higher profile.

This recognition of the importance of cross-sectoral approaches, environmental protection and technology transfer presents a major opportunity to energise the agricultural sector in developing countries. In pursuing this approach, developing countries have the opportunity to weaken the historical links between economic growth, energy consumption and environmental pollution. In the rural development sector, this can be achieved by relying more on sustainable energy technologies such as renewable energy systems and by using all fuels more efficiently through energy efficiency programmes. What is needed are integrated energy policies and frameworks, beyond project-specific environmental assessments, in which the costs and benefits of technology options can be assessed and compared on a sector-wide basis, allowing the right choices to be taken early in project design.

Income growth and industrialisation in developing countries are driving their economies towards the use of secondary energy carriers that deliver greater end-use energy efficiency. Typical of these carriers are electricity, distributed gas systems and liquid fuels. Modern biomass technologies can provide energy development pathways that meet all these requirements at the same time as offering both global environmental and local socio-economic benefits in rural areas.

Development of Technology

Historically most energy technology development has taken place in industrialised countries, where large, rapidly-growing demands created favourable conditions for innovation, and where commercial and industrial interests have found opportunities for profitable markets. In recent years, however, this growth has been confined largely to services and knowledge-based industries, which tend to require very few materials and energy resources. To the extent that innovation is still taking place in the energy sectors of industrialised countries, these are now driven by local and global environmental concerns.

On the other hand, rapid growth in energy producing and energy intensive activities in developing countries is expected to take place over the next 20-30 years. The conditions for innovation in energy technology are likely to be more suitable in developing countries, and hence these countries have an opportunity to stimulate techniques and technologies that could enable them to bypass some of the energy supply and demand patterns of the industrialised countries. This prospect of technological "leapfrogging" by developing countries could give them the chance to commercialise new technologies relatively quickly, provided there is a strong commercial incentive and a sustained market pull for such technologies. Rapid growth prospects in developing countries offer the opportunity for them to become market leaders in environmentally sound energy technologies.

The rural poor and landless require sustainable livelihood systems that are flexible and not over-dependent on a single resource or product. If large protected areas were managed for bioenergy feedstocks, local people might lose access to other products such as fibre and food. Bioenergy development policy must build in adequate provisions concerning local environmental and social factors, including means of achieving participation and in assessing and monitoring bioenergy activities.

For rural people to benefit directly from bioenergy projects, there need to be checks and balances to ensure that land use changes do not reduce equity. Flexibility to allow for local conditions, which might involve a mix of agroforestry, agriculture and dedicated bioenergy plantations. A robust strategy for developing bioenergy in a way that can assist local rural development should, therefore, include appropriate policies, institutions and community mechanisms.

Role of FAO

FAO has a longstanding global mandate from its member countries to promote renewable energy within the agricultural and forestry sectors. FAO has for many years stressed the potential of biomass as a locally available, renewable source of energy and has been active in implementing a range of field projects funded by donor agencies aimed at:

- — increasing the supply of biomass, reducing woodfuel consumption and improving the efficiency of end-use;
- — promoting renewable energy to enhance agricultural productivity and rural services;
- — addressing health problems and gender equality related to traditional woodfuel uses;
- — strengthening information and cooperation networks, including data collection and dissemination, awareness raising and capacity building;
- — promoting the adoption of biomass issues and actions into national agricultural, energy and environmental policies;
- — developing partnerships to take forward biomass activities with other bilateral and multilateral organisations, NGOs and the private sector;
- — increasing collaboration with agro-industries with biomass production potential.

FAO is currently strengthening its bioenergy programme with the aim of contributing to a partial substitution of fossil fuels by biofuels, as well as encouraging the more rational, efficient and safe utilisation of biomass.

Modern Bioenergy Technology

For the energy function of agriculture to contribute to the rural economy of developing countries, the potential role of energy production from biomass should receive greater recognition. There is also the need for positive political encouragement and appreciation of the social and cultural changes that might be needed to develop this potential. There is a balance to be struck between the advantages and disadvantages of bioenergy and this is reflected in both optimistic and pessimistic views regarding the uncertainties about how bioenergy systems can provide cost-effective local and global benefits. These uncertainties have restricted the development and commercialisation of modern biomass technologies. However, with the significant environmental pressures to redirect the global energy economy onto a more sustainable path, there is a real and urgent need to reconsider policy choices and commitments in the energy sector. Actions to take the energy sector in new directions are vitally important, and the proposed steps listed above are a first stage in this process, which is applicable to the rural energy sector of developing countries.

Policies and projects aimed directly at improving energy services for the poor are needed. One means of achieving this is by investments in pilot or demonstration projects that increase market uptake of promising new technologies, with active involvement of local people, NGOs and the private sector. For many developing countries, an energy transition would be characterised by a move from the present levels of subsistence energy usage based mainly on human work and woodfuel resources, to a situation where household, community and agricultural activities make use of a range of sustainable and diversified renewable energy resources.

Energy and agriculture in developing countries has long been considered as a secondary activity, with industrial and urban energy provision taking priority in many national energy programmes. Whilst no single energy source will have a monopoly of supply, it is clear that renewable energy sources in general, and bioenergy in particular, are likely to have larger role in future.

13

Forestry

The demand for forest products and services has increased in response to increase in population and income. Historically, production of wood and wood products has been the main objective of forest management and other functions were not explicitly accounted for. However, there has increasingly been a shift away from this towards assign'ing a higher priority to the environmental and social functions of forests. Conflicting demands and differing views of the relative importance of various goods and services provided by forests and alternative land uses will increasingly have to be reconciled. The complexity of predicting future developments in forestry is increased by demands for a more equitable distribution of the benefits from forests, for safeguarding the rights of forest dwellers and indigenous people, and for ensuring widespread participation in decision-making.

One of FAO's major forestry programmes is the periodic Forest Resources Assessment (FRA), which produces a statistical snapshot of the global status of forest resources once in a decade. The most recent of these, the Global Forest Resources Assessment 2000 estimates the global forest area in 2000 at about 3870 million ha or about 30 percent of the land area. Tropical and subtropical forests comprise 48 percent of the world's forests, while temperate and boreal forests account for the remaining 52 percent. Natural forests are estimated to constitute about 95 percent of global forests, while plantation forests constitute around 5 percent

Table 1. Global forestry at a glance, 2000

	Total forest				Annual cover change	
	Area	Share of land area	Area per capita	Plantation Area	1990-2000	
	'000 ha	%	ha	'000 ha	'000 ha	%
Africa	649866	21.8	0.85	8036	- 5262	- 0.78
Asia	547793	17.8	0.15	115847	- 364	- 0.07
Oceania	197623	23.3	6.58	2848	- 365	- 0.18
North and Central America	549304	25.7	1.15	17533	- 570	- 0.10
South America	885618	50.5	2.60	10455	- 3711	- 0.41
Europe	1039251	46.0	1.43	32015	881	0.08
World	3869455	29.6	0.65	186733	- 9391	- 0.22

Developing countries account for 2123 million ha (55 percent) of the world's forests, 1850 million ha of which in tropical developing countries. FRA 2000 revealed a net annual decline of the forest area worldwide of about 9.4 million ha between 1990 and 2000. Annual forest clearance was estimated at 14.6 million ha, and annual forest area increase at 5.2 million ha. Nearly all forest loss is occurring in the tropics.

Wood and Wood Production

Supplies of wood are not drawn on all global forests but from a significantly smaller area. FAO's Global Fibre

Supply Model estimates that only around 48 percent of global forests are available for wood supply. The remainder is either legally protected or physically inaccessible, or otherwise uneconomic for wood supply. The greater part of global wood production is burned as fuel; fuelwood at present accounts for over 1.7 billion m^3 or around 55 percent of the annual global wood harvest.

The vast majority of wood burning occurs in developing countries where wood is often the most important source of energy. Annual global production of industrial roundwood currently amounts to slightly over 1.5 billion m3, around 45 percent of the global wood harvest. Two-thirds of industrial wood products are consumed in developed countries, which account for less than one-fifth of the global population. Interestingly, annual per capita wood consumption in tropical and temperate and boreal regions is approximately equal (just over 0.5 m^3). In temperate and boreal countries, however, more than 75 percent of wood consumption is in the form of industrial wood products while, conversely, in tropical countries more than 85 percent of wood consumption is in the form of fuelwood.

Around 350 million of the world's poorest people are entirely dependent on forest ecosystems for their livelihoods. These people are often marginalised in terms of opportunities for development and, as such, are dependent on forests for their livelihood. Reducing forest pressures created by expansion of agriculture, grazing and fuelwood gathering depends on more general economic and social development to provide alternative income-earning opportunities. While development can reduce the pressures emanating from rural poverty, it also generates increased demands for a number of forest products. Direct and indirect impact of this on forests will largely depend on technological changes in production and the pattern of distribution of and access to forest products.

The global area of forests in formally protected areas is estimated to be 479 million ha, or 12.4 percent of total forest area. On the surface, this shows considerable progress towards the recommendation of the International Union for the Conservation of Nature and Natural Resources (IUCN) of having at least 10 percent of forest area in protected areas. Significant challenges still remain, however, in ensuring balanced representation of all forest types, and in ensuring effective protection of forests. At present, many countries still fall well short of having 10 percent of their forests protected, while in many other countries the representation of protected forest types is badly skewed. Furthermore, in many developing countries, protection is afforded to protected forests on paper only, while in reality they remain subject to substantial encroachment including logging, deliberate burning and other forms of clearing or degradation.

Shaping Forces of Forestry

Statements about what is likely to happen in the future necessarily include elements of speculation, increasingly so the longer the projection period. In many countries forestry is a low-priority sector and what happens to forests and forestry is to a great extent decided by changes outside the sector. The multifunctionality of forests, arising from the diverse values assigned to them, adds to the complexity of management. The development and future of individual forests, and of forests as a whole, will largely depend on the relative importance assigned to each of these values and on the strength of political advocacy for each value.

The key questions in developing an outlook for forestry over the next three decades are the factors that will have an impact on each of these demands and how policy environments will evolve to alleviate and reconcile demand pressures. Increasing demands for forest products arise from growing populations, increasing prosperity as a result of economic development and changes in consumer preferences. In this latter category, the most significant changes in preferences include the explicit recognition of the value of the environmental and social dimensions of forestry. Trends in each of these central, shaping pressures are examined below.

Demographic changes. Population growth, together with agricultural expansion and agricultural development programmes, is a major cause of forest cover changes. Population growth results in an increase in demand for food, which is met through intensification and arable land expansion. Changes in population distribution, especially urbanisation, will change the demand for fuelwood, especially charcoal, when incomes are insufficient to procure alternative sources of energy.

Depletion of wood resources from areas supplying urban centres has been a major problem in several countries and this is expected to persist during the next three decades, unless alternative sources of energy are more widely used. The overall impact of demographic changes on forests will vary between countries. In most developed countries where population has stabilised, deforestation has been arrested. Indeed the extent of forest cover is on the increase, primarily because of reversion of agriculture areas to forests. The number of countries where this process could happen is expected to increase.

Economic changes. Rising incomes and changes in income distribution are major factors affecting forestry. With rising incomes, demand for forest products increases, especially for processed items (e.g. panel products, furniture, printing and writing paper, and so on). In countries where control of access to forests is weak, increased demand for forest products is likely to exacerbate problems of poor forest management and excessive (including illegal) logging. In some countries, however, increasing prosperity may actually reduce some of the poverty-driven pressures on forests by providing alternative sources of livelihood. Growth of the industrial and services sectors reduces direct dependence on land.

Forests have often served as an engine for economic growth. In the early stages of economic development many countries have harvested forests as a means of capital accumulation. Most developed countries have gone through a phase of forest depletion during the early years of industrial development and many developing countries are currently moving through this phase. The challenge for developing countries that rely on forestry as a significant earner of foreign exchange is to maintain a healthy rate of development without excessively depleting forests.

The scenario that may emerge by 2015 and 2030 could be as follows:

— More countries in Asia and South America would have made significant progress in terms of economic growth, leading to significantly reduced dependence on land.

— Some countries that have depended on forests to generate investment funds would have diversified sufficiently, reducing their dependence on forests and increasing their willingness to adopt more sustainable management practices. However, there will still be several countries in Asia, Africa and South America where economic development and growth of the non-farm sectors would be inadequate to facilitate such a major shift from land-based activities. Deforestation and its attendant problems are likely to persist in these countries.

— In several countries economic progress would enable substantial increases in investment in large infrastructure projects, often resulting in deforestation. This is particularly the case in a number of countries in Asia and South America. Forest clearance for infrastructure development (such as large irrigation and hydroelectric projects) and expansion of area under industrial crops could counterbalance the reduction in forest clearance for arable cropping.

— Even in Asian and Latin American countries that make significant economic progress, if large areas of deprivation persist, people will continue to depend on land and other natural resources, exerting direct and indirect pressure on forests.

— While consumption of forest products is expected to stabilise in the developed economies, an increase in consumption of all wood products is expected in most developing countries.

— The demand for forest services such as recreation and ecotourism is expected to increase.

Political, social and institutional changes. The last two decades have witnessed major changes in the openness, transparency and decentralisation of governance. Local communities and groups have achieved an increasing say in the management of natural resources.

More pluralistic institutional arrangements for resource management are expected to emerge and strengthen. Several countries have revised their forest policies and changed legislation, enabling the participation of the private sector in forest management, including community groups and farmers. Participatory approaches such as joint forest management and management by forest-user groups are becoming more acceptable, although the area managed under such arrangements is still very small. These efforts are expected to gain momentum during the next three decades.

Forestry departments in many developing countries could change their structure and functions. Their primary responsibility will shift from direct management of forests to policy development and other regulatory functions. Actual resource management will largely be the responsibility of the private sector, including farmers and local communities. The proliferation of information (and the potential for misinterpretation) requires proper interpretation and improved public access to information. This will be another important emerging function of the public sector forestry agencies.

Forests, environment and global initiatives. The growing concern about the environment encompasses a wide range of forest functions, including protection of watersheds, conservation of biodiversity and the perceived role of forests as carbon sinks. As water becomes a more critical resource, the role of forests in flood control, in rainfall interception and in controlling erosion will assume even greater importance, requiring better knowledge and quantification.

These concerns have become a key driving force in forestry initiatives following the 1992 UN Conference on Environment and Development (UNCED). UNCED Agenda 21, in particular Chapter 11, has formed the basis for several initiatives. Three international, legally binding conventions, the Convention on Biodiversity (CBD), the

UN Framework Convention on Climate Change (UNFCCC), and the Convention to Combat Desertification in those Countries experiencing Serious Drought and/or Desertification, particularly in Africa (CCD), are of direct relevance to forestry. They have highlighted the importance of forests in providing a range of goods and environmental and social services, and have helped to widen the focus of international attention from forests proper to trees in fragile ecosystems such as drylands and mountains.

National and international non-governmental organisations (NGOs) and advocacy groups have brought forestry and environmental issues to the forefront of public debate. NGOs and civil society will continue to have a pioneering role in influencing forestry policies and programmes. Since UNCED, countries have attempted to make forestry policy more consistent with the concepts of sustainability and environmental soundness. The Ad hoc Intergovernmental Panel on Forests (IPF), the Intergovernmental Forum on Forests (IFF) and the United Nations Forum on Forests (UNFF) have helped to articulate the policy, institutional, social, environmental and technological issues relating to forest management, although commitment to a legally binding mechanism for conserving and managing forests still remains elusive.

These developments have not only generated pressures against the conversion of forests to alternative land uses but they have also resulted in a closer scrutiny of forestry practices, including logging of natural forests and plantation management. These concerns are expected to dominate the debates at local, national and global level and would have the following implications:

— increasing concern for protecting the environment, leading to a better recognition of the global, national and local significance of forests; improved mechanisms to determine the trade-offs between competing objectives; and appropriate arrangements to share costs and benefits;

- recognition of the rights of indigenous and forest-dependent people and an increasing concern to protect their livelihood;
- integration of tree cultivation with other land uses, to reduce environmental problems and to enhance income opportunities; and
- global and regional initiatives to enhance the role of forests in mitigating climate change, especially their potential as carbon sinks.

International trade. A significant proportion of global production enters international trade, reflecting the difference between the geographic distribution of industrial roundwood production and processing capacity, and of demand for forest products. Table 2 shows that the export share of more processed products is significantly higher than that of less-processed products, reflecting most countries' efforts to capture foreign exchange earnings through higher value-added forest processing. For example, the export share of sawnwood and wood-based panels is much higher than that of its raw material, namely industrial roundwood, and likewise for paper and wood pulp.

The largest forest product trade flows are intra-regional. Europe, North America and Asia combined account for 90 percent of the value of forest product exports and around 95 percent of imports. Europe accounts for almost a half of the global trade in forest products, and 80 percent of European trade is between European countries. Similarly, trade between the United States and Canada accounts for around 70 percent of

North and Central American trade, while Japan is a partner in around 35 percent of trade in the Asia-Pacific region.

Table 2. Trade in forest products as a proportion of production, 2000

Region	Industrial roundwood			Sawn wood			Wood-based panels			Pulp			Paper and paperboard		
	P	E	S	P	E	S	P	E	S	P	E	S	P	E	S
	mln	m^3	%	mln	m^3	%	mln	m^3	%	mln	mt	%	mln	mt	%
Africa	69	6	9	8	2	25	2	1	50	3	1	33	3	1	33
Asia	229	11	5	54	7	13	47	15	32	22	2	9	95	12	13
Oceania	47	9	19	8	2	25	4	1	25	3	1	33	4	1	25
North and Central America	624	15	2	194	57	29	60	14	23	86	18	21	112	25	22
South America	153	2	1	30	5	17	11	3	27	11	5	45	10	1	10
Europe	464	71	15	128	56	44	65	23	35	47	11	23	100	58	58
World	1587	114	7	421	128	30	189	56	30	171	38	22	323	98	30

Note: P=production; E=exports; S=exports as share of production (in percentage).

As further trade liberalisation takes place and barriers to trade diminish, there could be significant changes in the composition, volume and direction of trade during the next three decades. Countries will continue to develop domestic wood-processing facilities, and thus the share in trade of processed and semi-processed products will increase relative to that of unprocessed products. Economic growth in Asia is creating new opportunities for trade of processed and unprocessed wood and this could accelerate the shift in the direction of trade already evident. Increased demand from Asia coupled with restrictions or outright bans on logging of natural forests in some of the Asian countries have to some extent contributed to unsustainable (often illegal) logging in Africa and Latin America.

Developments in science and technology. Developments in science and technology and their wider diffusion could lead to the emergence of more knowledge-based operations and management. The ability to take advantage of natural processes will increase considerably, leading to more benign approaches to resource management. Access to knowledge and information will improve, and even societies that have historically been isolated will have access to information. Of particular interest to forestry are the following probable developments:

- Better understanding of ecological processes will improve the ability to fine-tune human interventions, minimising the adverse effects of several forestry practices.
- Application of improved remote sensing technologies will allow the state of resource changes to be assessed on a real-time basis and to be available to wider groups of people.
- Improved knowledge of the chemical and genetic make-up of a large number of species will widen the scope for their use, leading to the emergence of new processes and products.
- Significant improvements in processing technologies will reduce raw material requirements and environmental pollution.
- Increased use of tree breeding and application of vegetative propagation technologies will ensure

availability of high-quality planting materials, especially for establishment of large-scale plantations.

— Energy sources will probably shift, with wind and solar power becoming more important and widely accessible. This may have a significant impact on forestry, especially on the use of fuelwood.

Tree breeding is increasingly drawing on a range of biotechnological tools. These fall broadly into three categories: (i) biotechnologies based on molecular markers that help quantify genetic diversity and identify promising genotypes and genes for subsequent use in breeding programmes; (ii) technologies that enhance vegetative propagation; and (iii) genetic modification. While the first two techniques are relatively commonly used to support existing tree-breeding programmes, genetic modification is still at an experimental stage. To be of practical value, genetic modification must offer unique features that cannot be economically delivered through selection or more conventional breeding and that are capable of offsetting the costs and time involved.

All new technologies must be tested and proved to be environmentally safe before they can be used on a large scale. Important traits in forest trees, such as growth rate, adaptability to harsh environmental conditions and stem and wood quality, are governed by an array of genes and may not lend themselves readily to genetic modification. Should genetic modification be pursued on a large scale in forestry, the characteristics most likely to be targeted are simple inherited traits, such as modified lignin content, insect and virus resistance and herbicide tolerance. There are, to date, no reported large-scale or commercial plantations of genetically modified trees.

Although changes up to 2015 may be slow, by 2030 developments in science and technology are expected to have more far-reaching impacts. But the pace of adoption of technological change will vary enormously between and within countries. This will depend on investment in human resources, on providing favourable institutional arrangements, etc. In the short term, disparities in access to and ability to use knowledge may widen. Large areas of underdevelopment may persist in a number of countries, limiting the participation of a sizeable section of people in knowledge and technology-based activities.

Globalisation and its impacts. Economic liberalisation and increasing interdependence will provide new opportunities, but will also mean that perturbations in one sector or country could easily affect other sectors and countries. Environmental interdependence such as in global climate change or the transboundary effects of deforestation and other related changes, could affect economic interdependence. The implications of economic and environmental integration on forestry could be far-reaching. Of particular importance could be the following factors:

— The movement of technologies, processes and products across countries will increasingly be based on comparative (competitive) advantages. Traditional forest-based enterprises that fail to take advantage of the emerging opportunities will fade out and be replaced by more efficient producers.

— Regional and international trade agreements will continue to alter the pattern of trade in forest products. While tariff barriers may become less important, several non-tariff barriers, such as those relating to certification and ecolabelling, could become critical in determining access to certain markets.

— Global, national and local pressure groups will play an increasing role in shaping forest management, each group trying to push their agenda based on the group's interests and objectives.

However, economic integration is not necessarily a benign process conferring benefits to all at the same time. Disparities in access to knowledge, resources and markets could widen the economic gap between countries, and could also accentuate resource-use conflicts within countries.

Future Changes

Forest and tree resources. The key issue for forest and tree resources is progress towards wider adoption of sustainable forest management (SFM). There is some evidence of progress, although scattered and disparate. In some areas, excellent forest management complies with criteria agreed upon for SFM. Elsewhere, mainly in the tropics, substantial tracts of forests remain unmanaged and are frequently severely degraded by careless logging, burning or other destructive practices. Initiatives to remedy the situation include the development and implementation of criteria and indicators for SFM, and subsequent adjustments to forest management practices; establishment of certification processes; implementation of pilot projects and operations aimed at reduced impact logging; and extension of improved silvicultural systems.

All these initiatives are helping efforts to manage forests to meet a range of long-term objectives and demands. In most developed countries, forest stocks have been increasing for a number of years, both because management of existing forests has been intensified (including the establishment of forest plantations) and because significant land areas of marginal farmland have been allowed to regenerate as secondary natural forest. The challenge for these countries is to continue to improve forest management to encompass the new dimensions of sustainability. Conversely, in a number of developing countries, even to arrest deforestation and forest degradation would mean considerable progress.

Shift in the source of wood supplies. During the next three decades, the shift away from dependence on natural forests for wood production is expected to strengthen. Increasingly wood is being obtained from intensively managed plantations, thus reducing the pressure on natural forests and woodlands and enabling the latter to fulfil other functions. Between 1980 and 1990 the area under forest plantations increased from about 18 million ha to nearly 44 million ha. By 2000 it had increased further to a total of 187 million ha. Although plantations account for only 5 percent of forest cover, they have become a significant source of roundwood supply.

There is also an increase in tree cultivation on farms and other land outside forests. In several countries (e.g. in Bangladesh, India, Kenya and South Africa) trees from farms are an important source of industrial roundwood and fuelwood. Often such efforts are supported by industries through provision of technical assistance and assured prices. Stable or increasing prices and secure tenure are needed to encourage such efforts. As farmers adopt more integrated land uses with trees as an important component, an increasing proportion of local wood supplies is expected to come from trees outside forests.

Consumption of industrial roundwood. Consumption of industrial roundwood during the next 30 years is expected to move modestly upwards. The most recent projections of FAO's Global Forest Products Model estimate that global consumption of industrial roundwood in 2030 will be around 2400 million m^3, an increase of around 60 percent on current consumption. Much of this increase will result from an increase in population and income in developing countries, leading to an increase in per capita consumption of wood (and particularly paper) products. Offsetting these trends will be changes in consumer preferences, changes in techniques of forest management, more efficient utilisation of wood and residues, and the development of new products.

Current indications are that the real prices of most forest products are unlikely to increase significantly in the long term. There may be short-term fluctuations, but these will be evened out through changes in supplies and shifts in sources of supplies, including substitution by non-wood products. Similarly, the spread of SFM may reduce supply and increase costs in the short term, but it is unlikely to affect supplies in the long term. In fact, by making supplies sustainable, it is expected to stabilise long-term supplies and thus help maintain price stability.

A marked change will take place in the share of roundwood grown on plantations. Industrial roundwood production from plantations is expected to at least double during the next 30 years, from the current 400 to around 800 million m3. By 2030 plantations will supply about a third of all industrial roundwood. Much of the growth in demand for wood during this period will, consequently, be met from increased plantation supplies, and this will necessitate some technological changes to match wood qualities with consumer demands.

In parallel, significant changes in the type of timber produced in natural forests are likely to induce shifts in the relative prices of forest products, leading to shifts in demand. For example, there is an increasing scarcity of large tropical logs, which will necessarily constrain the production of hardwood plywoods and large-dimension sawn timber. Consequently, demand for softwood, plywood or other reconstituted panels may increase. Similarly, solid wood products made from tropical timbers are likely to become premium products leading to a shift in demand towards cheaper wood-veneered products.

Changes in forest-based industries. Changes in prices and demand, together with the development of new technologies and changes in resource availability, are affecting the types and volumes of forest products. The most striking change in the manufacturing of forest products during the past 30 years has been the growth in production of wood-based panels and paper products as compared with sawn timber. Since 1970, global production of sawn timber has remained largely static, while production and consumption of wood-based panels have more than doubled and production of paper and paperboard has almost tripled.

As Table 3 shows, output levels of processed wood products (wood-based panels and paper products) have been increasing, even though levels of industrial roundwood production have remained static or declined. This has been made possible by improvements in processing efficiencies. This trend encompasses the effects of utilising recycled paper and wood but, more important, the development of products that minimise waste by making fuller use of processing residues. A generation ago, solid wood furniture was the norm, but today much utilitarian furniture is made from fibreboard with a veneer covering or from cheap and abundant softwoods such as plantation-grown pines. Similarly, there has been considerable expansion in the types and range of paper products. The development of a broader range of reconstituted paper and panel and board products has also allowed much greater use of processing residues, enhancing the overall efficiency of wood use.

Table 3. Average annual production of selected forest products

Five-year period	Industrial roundwood	Sawn wood	Wood-based panels	Wood pulp	Paper and paper board	Recovered paper
	mln m3	mln m3	mln m3	mln mt	mln mt	mln mt
1971-75	1318	428	87	102	131	35
1976-80	1415	452	102	125	170	51
1981-85	1457	445	104	136	193	61
1986-90	1660	503	124	155	240	83
1991-95	1501	437	130	162	282	104
1996-00	1523	419	165	171	323	134

One of the key changes that can be expected over the period to 2015 is the rationalisation of industry capacities. At present there is a significant global overcapacity in pulp and paper facilities, panel plants and sawmills, as evidenced by relatively low returns to forestry sector investments. Much of this overcapacity is the result of distortions in investment decisions created by government incentives (including low stumpage prices, direct investment incentives and tariff protection). Obsolete processing technology and inappropriate location of processing have increased the uncompetitiveness of forest-processing industries in many countries.

A number of factors are forcing greater competition in forest product markets. Trade liberalisation is an obvious one. The pressures created by wood scarcity, widespread adoption of market-based economies and greater environmental scrutiny are other factors altering the structure and location of forest-based industries. The period to 2015 and 2030 will certainly witness a much greater reliance on competitive advantages in determining patterns of forestry industry investment.

Wood as a source of energy. As noted above, an estimated 55 percent of global wood production is used as fuelwood. Tropical countries account for more than 80 percent of global fuelwood consumption. There is, however, considerable regional and national variation. Asia and Africa together consume more than 75 percent of global fuelwood. The Regional Wood Energy Development Programme (RWEDP) in Asia estimates that wood supplies 18 percent of all energy used in its member countries. In Africa wood is still the most important source of household energy, used mainly for cooking, although cottage industries (such as food drying and brick-making) also consume significant volumes in some countries.

Fuelwood consumption is mainly determined by income levels, availability of wood supplies and availability of alternatives. As consumers become wealthier they prefer other forms of conventional energy, particularly electricity and liquid fuels. Therefore, while overall consumption of fuelwood is increasing, per capita consumption is declining. In most countries there is a surplus of fuelwood, although there is often localised scarcity, especially in areas adjacent to urban centres, which are sometimes subject to rapid deforestation for charcoal production. In some cases, increasing electricity and gas prices (for example because of the privatisation of utilities) have resulted in a switch back to fuelwood.

Dramatic changes in fuelwood consumption are unlikely over the next 15 years. Access to alternative fuels will become easier, but the majority of current wood-using communities are likely to be still burning wood in 2015. The shift towards alternative fuels may accelerate beyond 2015, depending on developments of infrastructure and on improvements in the efficiency and cost-effectiveness of generating energy.

Role of forests in mitigating climate change. Increasing concern over CO_2 emissions and their potential impacts on global climate has focused attention on the role of forests in regulating atmospheric concentrations of carbon. Forests serve as carbon reservoirs by storing large amounts of carbon in trees, understorey vegetation, litter and soil. Newly planted forests, or degraded forest areas that are allowed to regenerate, act as carbon sinks by absorbing and storing carbon as their biomass increases. Conversely, when forests are cleared or degraded, their sink potential is reduced and they can become a substantial source of CO_2 emissions. Therefore projects that focus on afforestation, improved forest management and forest protection can reduce net volumes of atmospheric carbon dioxide and thus play an important role in mitigating climate change.

Estimates of the carbon sink capacity of forests suggest a potentially highly significant role. Overall, forests contain just over half of the carbon in terrestrial vegetation and soil, amounting to 1200 Gt of carbon. Boreal forests account for more carbon than any other terrestrial ecosystem (26 percent), while tropical and temperate forests account for 20 and 7 percent, respectively. The Intergovernmental Panel on Climate Change (IPCC) has estimated that, globally, carbon sequestration from reduced deforestation, forest regeneration and plantation development could equal 12 to 15 percent of the total carbon dioxide emissions expected to be generated by fossil fuels between 1995 and 2050.

The Kyoto Protocol is the key framework for international agreement on stabilisation of greenhouse gas concentrations in the atmosphere. Article 3.3 of the

Kyoto Protocol clearly states that "the net change in greenhouse gas emissions by sources and removals by sinks resulting from direct human-induced land-use change and forestry activities, limited to afforestation, reforestation and deforestation since 1990, measured as verifiable changes in carbon stocks, shall be used to meet the commitments under this Article of each party". However, the protocol is unclear about the extent to which forests will be included in the principal mechanisms for regulating carbon emissions. There remains considerable debate over rules for incorporating forestry projects in emission reductions under the clean development mechanism (CDM). Some efforts are under way to develop trading in carbon emission permits and some CDM pilot projects are under implementation, but there is still uncertainty regarding the future potential of these arrangements.

Non-wood forest products (NWFPs). Forests produce many other products besides wood. The vast majority of these NWFPs are subsistence goods, used by the household that collects them, or traded only in local markets (Table 4). There are, however, a number of products that have been developed commercially and an estimated 150 NWFPs are traded internationally. Most are traded in relatively small quantities and supplies are typically unstable, with varying quality and unreliable supply chains. There are, however, a number of exceptions to these general trends including, for example, rattan, bamboo and gum Arabic.

As mass-produced consumer goods become increasingly accessible to even very remote communities, the dependence on inferior NWFPs has declined. The disintegration of many traditional livelihood systems has exacerbated this trend, as has the overexploitation of many NWFP stocks. In the past, the commercialisation of NWFPs has been viewed as a potential development path for forest dwellers in less developed countries. Increasingly, however, there is recognition that natural systems for the production of NWFPs are not intensive enough and generally too unreliable to provide a basis for large-scale industrial development.

Table 4. Classification of non-wood forest products (NWFPs) by end uses

End use	NWFP examples
Food products and additives	Wild meat, edible nuts, fruits, honey, bamboo shoots, birds' nests, oilseeds, mushrooms, palm sugar and starch, spices, culinary herbs, food colorants, gums, caterpillars and insects, fungi
Ornamental plants and parts of plants	Wild orchids, bulbs, cycads, palms, tree ferns, succulent plants, carnivorous plants
Animals and animal products	Plumes, pelts, cage birds, butterflies, lac, cochineal dye, cocoons, beeswax, snake venom
Non-wood construction materials	Bamboo, rattan, grass, palm, leaves, bark fibres
Bio-organic chemicals	Phytopharmaceuticals, aromatic chemicals and flavours, fragrances, agrochemicals/insecticides, biodiesel, tans, colours, dyes

Development of NWFP programmes is also hindered by lack of focus, institutional and policy neglect, lack of research and development funding by NWFP producers and lack of capital. The future is likely to see only a handful of key species being cultivated on a commercial scale no longer controlled by forest dwellers. Thus, the opportunities for poor forest-based communities to profit substantially from NWFP development are likely to remain very limited.

Protected areas, biodiversity and environmental protection. IUCN has established a goal of having 10 percent of each country's land area under some form of protected status. Ideally, this should include a representative sample of all ecofloristic zones, so that 10 percent of all types of natural forests would be protected. At present, it is estimated that some 80 countries have formally attained the IUCN 10 percent goal, but around 100 countries currently still have less than 5 percent of their national land area under protection. Overall, it is estimated that more than 30 thousand protected areas have been established and that these cover about 3.7 percent of the global land area.

Working in collaboration with FAO's Global Forest Resources Assessment 2000, the World Conservation Monitoring Centre (WCMC) reported that the number and extent of protected areas grew steadily throughout the latter part of the twentieth century. However, much work needs to be done to harmonise national and international data, including data from different agencies in the same country. In addition, the interpretation of the concept of protected areas often differs substantially among countries, making aggregation at the global level unreliable.

Although the number and extent of protected areas have increased markedly in most regions since the 1960s, in many countries there will be diminishing opportunities to establish protected areas in their traditional form. In countries where conservation efforts fall below the IUCN target, there are often already intense land use pressures, or strong conflicts between economic and environmental objectives. Consequently, there will be a need to develop innovative means to combine conservation objectives with SFM and the creation of sustainable livelihoods. For the next 30 years, therefore, it is expected that the total land area under strict protection will increase only moderately, but greater areas are likely to be placed under SFM, in which conservation concerns are among the criteria.

Biodiversity is important at the level of ecosystems, species, populations, individuals and genes. There is a need for better management to safeguard biodiversity at all these interacting levels. While some losses in biodiversity over time are inevitable through both natural and human-induced causes, diversity can be conserved and managed through a wide range of actions, from the establishment of nature reserves and managed resource areas, to the inclusion of conservation considerations in improvement and breeding strategies of species under intensive use. The key to success will lie in the development of programmes that include a strong element of active gene management and that harmonise conservation and sustainable use of forest genetic resources within a mosaic of land use options. The sustainability of programmes will depend on genuine efforts to meet the needs and aspirations of all interested parties, and will require close and continuing collaboration, dialogue and involvement of stakeholders in planning and executing related programmes.

Some environmental threats will continue to require protective actions involving forestry. For example, combating desertification will remain a priority for many sub-Saharan countries, as well as throughout the Near East, and parts of North and South Asia. Watershed protection and flood control will continue to be crucial in countries with extensive steep lands and in their downstream neighbours. Weaknesses in institutional arrangements to mobilise concerted action will remain the principal constraint, exacerbated by the lack of financial resources, and difficulties in identifying beneficiaries and their willingness and ability to pay.

Forest-based recreation and ecotourism. Ecotourism is a rapidly growing function of forests. It is defined as ecologically sustainable tourism that fosters employment and cultural understanding, appreciation and conservation. More generally, it comprises a symbiotic relationship between the environment and tourism, and provides an economic incentive for conservation. Ecotourism comprises a relatively small part (about 7 percent) of the world's travel and tourism industry. Nonetheless, it constitutes a significant source of foreign exchange earnings in some countries. A major question mark is the extent to which it can serve as a means of financing conservation.

In developed countries, where leisure activities are most extensive, recreation markets are becoming more competitive. It is thus quite conceivable that actual forest-based recreation may decline over the next 30 years, even though forests are likely to be increasingly accessible. If increased recreation does occur, it is most likely to be where significant other services are provided. In developing countries, forest recreation is likely to increase together with increased affluence and the development of infrastructure to facilitate access to forests.

Food security and poverty alleviation. There is increasing discussion about what forests and forestry could do to alleviate the widespread prevalence of poverty and food insecurity. In many countries, forests and forest products are an important source of livelihood for rural communities. Traditional approaches to forest management by public sector agencies or the private sector have failed to consider this aspect. In recent years poverty alleviation has become the focal area of attention for international organisations, and the role of forests and forestry in this can be expected to be reconsidered.

Future Perspectives

Emerging markets and their impacts on the direction and volume of trade. One of the principal questions relating to the future development of industrial forestry is the extent to which new markets will emerge, and the effects they may have on directions and volumes of trade. Several demand growth poles may emerge over the coming 30 years, with significant impacts on global trade. The largest potential markets are in the populous countries of Asia, particularly China and India. At present these two countries, with around 40 percent of world population, consume less than 10 percent of the world's industrial wood. Other developing countries such as Indonesia, Brazil, Bangladesh and Nigeria all have large populations, but relatively low per capita rates of industrial wood consumption. If developing countries like these significantly raise their per capita consumption of industrial wood products, there will be a significant shift in patterns of trade.

Impact of globalisation

Continuing globalisation has an impact on markets for forestry and forest products in four ways: further trade liberalisation in forest product markets; standardisation in production processes, product quality and regulatory environments; an increasing role for foreign investment and investment by multinational corporations; and wider spread of environmental advocacy and concerns. During the past decades there has been significant liberalisation of trade in forest products. The liberalisation process culminated in the General Agreement on Tariffs and Trade (GATT) Uruguay Round, during which the majority of tariff barriers for forest products were reduced to fairly moderate levels or completely removed. Similarly, the past two decades have seen considerable standardisation in market and regulatory requirements.

Most countries have adopted trade, investment and even environmental policies, which have reduced national insulation and promoted a greater level of standardisation. Thus, while there is still room for further progress towards free trade in forest products and standardisation in markets, changes in trade conditions during the next 30 years are likely to be less important than other factors in determining the shape of forestry.

The globalisation of environmental advocacy really came of age at UNCED in 1992. UNCED witnessed the birth of global acceptance of sustainable development, and acceptance of SFM as the core paradigm for forestry development. At the same time, throughout the 1990s environmental advocacy has firmly embraced the concept of transnational environmentalism. Issues such as global warming and the role of forests in carbon sequestration, the development of multicountry criteria and indicators of SFM, and international forest certification processes are all indicative of growing concerns about the environmental implications of forestry. Perhaps even more important implications are those related to the globalisation of environmental concerns.

The scrutiny and interventions of environmental NGOs and development agencies have imposed greater accountability on multinationals and have forced governments to write and enforce environmental legislation and regulations. Governmental decisions relating to forests are now subjected to wider scrutiny at local, national and global levels, and there are increasing pressures to make forestry practices compliant with widely acceptable environmental standards. There is a long history of foreign direct investment (FDI) in forestry operations.

Most FDI in forests and forest-processing operations is undertaken by multinational companies. FDI is seen as a means of developing national economies by generating employment, income and foreign exchange earnings. Education, training and technology transfer also accrue to the "host" country. FDI in forestry operations, particularly those relating to forest purchases or harvesting concessions, is often controversial. Forests are often seen as strategic national assets, and there has been considerable resistance in many countries to relinquishing control of forests to foreign interests. Similarly, multinational companies operating in countries with weak regulations have been responsible for much destructive forest harvesting. Multinational companies are often perceived to be less accountable for environmental damage than domestic companies or government logging agencies.

Privatisation of forests remains a significant vehicle for promoting FDI in forests although, with the exception of a few countries, it has not progressed very far. It seems likely that privatisation will eventually progress in a number of countries, although this process is likely to be slow and is more likely to find favour in developed countries. Eastern Europe, where there are efforts to transfer formerly centrally controlled forest assets to private ownership, may be an exception to this rule.

Impact of developments in science on forestry practices. In simple terms, a central objective of most forestry science is to improve some part of the processes that convert sunlight into cellulose or of those that deliver forest goods and services. Traditionally, forestry research focused on making the production process faster, cheaper and better. Such research focused on seedling production, tree growing, harvesting and wood processing as well as all the intermediate steps. The bulk of forest science still operates within these parameters, and over the coming 30 years we should expect improvements across the broad range of the production process, making wood production faster and delivering cheaper and better forest products.

An equally important new trend is to incorporate social and environmental objectives in forest science, management and production. In a number of countries paper production is required to include a minimum proportion of recycled material. Similarly, intensive forest harvesting covering large areas is increasingly considered unacceptable by the public, even though in many instances it is (in the short term) the cheapest means of harvesting and may mimic nature better than alternative systems.

There are also important changes in the technology supporting the monitoring of forests. The increased ease in collecting information (through satellite imagery, the Geographic Information System [GIS], etc.), collating it (database technology), and disseminating it (e-mail and the Internet) means that countries' forest management practices are under far greater international scrutiny. Thus there is far greater leverage to ensure that forestry practices are sustainable.

Deforestation and Forest Degradation

Deforestation refers to change of forest land cover where tree crown cover is reduced to less than 10 percent. Deforestation has the highest profile of all environmental challenges confronting forestry. It has an impact on the availability of forest goods and services by reducing the area of forest. High rates of deforestation (and forest degradation) may also increase pressure on forestry policy-makers to place forest areas in strictly protected areas or to prevent harvesting in other ways (e.g. through logging bans). This may increase pressure on accessible areas and further reduce production that can be harvested from remaining forest areas.

There is an identifiable relationship between deforestation and levels of industrial development. In subsistence economies with slowly growing populations, forests may absorb the impact of demands for wood and other forest products. The advent of widespread economic development and rapid population growth has generally seen a period of rapid deforestation in industrialising countries. At a certain level of development, however, further land clearing and forest harvesting become uneconomic, and may also generate increasing criticism related to neglect of social and environmental dimensions of forest management. Thus there exists an apparent curve of changing forest use, which leads from subsistence, through heavy exploitation, to harmonisation of harvesting with protection and conservation, as well as social needs and aspirations.

Forest degradation refers to changes between forest classes (e.g. from closed to open forest) that negatively affect the stand or site and, in particular, lower production capacity. Extensive logging, without corresponding attention to forest management, has led to large-scale forest degradation, especially where the extraction of mature trees is not accompanied by their regeneration, and where the use of heavy machinery causes soil compaction or loss of productive forest area. Other causes include overgrasing, overexploitation (for firewood or timber), repeated fires and attacks by insects, diseases, plant parasites or other natural sources such as cyclones.

Forest degradation takes different forms. In most cases, degradation does not imply a decrease in the area of woody vegetation, but rather a change in species composition, loss of soil fertility, erosion, reduction in biodiversity and fragmentation of forest areas. One of the most easily identifiable and important indicators of degradation is depletion of standing biomass, particularly mature trees. Forests may be degraded from virgin primary status to secondary forests, to scrub or other wooded land and then perhaps cleared and reafforested as plantation. Broad measures of forest area and deforestation do not capture the changes in values that these shifts encompass.

14

World Fisheries

Modern technology has changed the nature of the fisheries. For centuries the industry revolved around codfish, caught in inshore waters and salted or dried primarily by small family-run facilities for domestic and international markets. Depleted fish stocks, notably of cod, have reduced catches overall, and shellfish have become more important. Secondary processing is assuming an increasingly important role. Agricultural products, consisting mostly of vegetables, fruits, and livestock, are marketed locally, and most of the foodstuffs consumed must be imported.

During the 1990s, the production of capture fisheries fluctuated between 80 and 85 million tonnes p.a. for the marine sector. Fish from the Pacific Ocean dominated world capture fisheries, accounting for almost two-thirds of total world supplies in 1999. During the same period, the production from fisheries in inland waters expanded slightly, increasing from 6.4 million tonnes in 1990 to nearly 8.3 million tonnes in 1999. However, inland fishery catches are believed to be greatly under-reported. This is because of the dispersed and informal nature of many inland fisheries, and because many inland fish are bartered, sold or consumed locally without entering into the formal economy.

For example, inland fisheries in Brazil, Ghana and several Southeast Asian countries may be twice to six times as productive, in terms of catch actually taken, as officially reported. In several countries efforts are under way to improve the data on inland fisheries and on the range of services inland waters provide. One preliminary result is that it appears to be more profitable to use Brazilian floodplains for fisheries than to use cleared floodplains for cattle grazing.

The gross volume of marine catches has fluctuated and does not show any definite trend. But the species composition of the catch has changed. High-value species—bottom-dwelling species (demersal) and large surface-dwelling species (pelagics)—are gradually being substituted by shorter-lived pelagic and schooling fish. FAO studies indicate many causes for this shift. These include the thinning out of (overfished) top predators; increases in natural production of small pelagics through nutrient enrichment of coastal areas, enclosed and semi-enclosed seas; and changes in fishing strategy and technology. The main force underlying such changes, however, is the change in harvesting costs as fishing technology has advanced and as various stocks have been depleted, and the impacts of these changing costs on operations.

During the second half of the twentieth century, capture technologies evolved rapidly. In the 1950s the invention of gear handling devices, such as the power block for purse seines, resulted in major improvements in the efficiency of fishing methods such as trawling, gillnetting, purse seining and lining. The development and introduction of synthetic fibres in fishing gear during the 1960s significantly increased the efficiency of such gears, thereby lowering that cost component of production. Also, new vessels were able to remain for longer periods in the fishing grounds and the introduction of freezing on board helped increase the autonomy of vessels.

Modern fishing fleets have also seen continued growth in efficiency over the last decades, thanks to improvements in navigation and fish-locating equipment, increased knowledge of fish behaviour during the capture process, and development of gear monitoring instruments. The growth of efficiency has contributed to increased exploitation rates and overexploitation in some fisheries, as well as to overcapacity in others.

A more general assessment of the condition of world stocks depends on how one categorises the stocks that at present are exploited close to the maximum sustainable yield (MSY) level. If the fully exploited stocks are accepted as being in good condition because they comply with the United Nations Convention on the Law of the Sea (UNCLOS) requirement of being at or above the MSY level of abundance, then 72 percent of the stocks appear exploited at sustainable levels. However, this still implies that 28 percent of stocks are overfished or even depleted and therefore need rebuilding.

Until the 1950s, aquaculture involved a limited number of species: oysters, mussels, carp, trout and shrimps. This was largely because these were the species for which large volumes of young/spat could be easily obtained for production purposes, and such culture systems were generally extensive. Since the 1950s, scientists gradually solved the problem of artificial reproduction for different carp (e.g. common, Chinese and Indian carp), salmonids, oysters and shrimps, making it possible to stabilise and expand production of these species.

The bulk of aquaculture production comes from species that feed low on the food chain. Among the finfishes, carp, which generally feed on plankton and plants, accounted for nearly 70 percent of world production of cultured finfishes in 1999. Commanding low to moderate prices, such fish tend not to be traded internationally. Aquaculture currently provides a considerable share of the production of several high-priced species such as salmon, marine shrimps and oysters.

In general, efforts to culture other finfish and crustaceans implied protracted and costly research and development efforts. This has hampered greater aquaculture-based production of many of these types of species. As a result, even by the end of the 1990s, a relatively small number of species (less than 40) accounted for 90 percent of the volume of world aquaculture production.

During the 1990s, aquaculture production increased at the rapid rate of approximately 10 percent p.a., rising from a 1990 level of 13 percent of world fisheries production (excluding aquatic plants) to 26 percent nine years later. Currently, aquaculture production already far exceeds the year 2010 production level projected as recently as in 1995. Aquaculture's share of fish used in food consumption has risen in volume terms from 19 percent in 1990 to 34 percent in 1999. In value terms, this contribution is 39 percent—the ex-farm value of aquaculture (fish and shellfish) production was estimated in 1999 at approximately US$48 billion, compared with an estimated landed value of capture fisheries at approximately US$76 billion.

The growth of aquaculture has been unevenly distributed around the globe. Production has been heavily concentrated in Asia both in volume and value terms. At the beginning of the 1990s, Asia provided 83 percent of the world's total volume of aquaculture production, and by 1999 that share had risen to 89 percent. Most of the increase was attributable to China, whose share of the world's aquaculture production increased from 50 percent in 1990 to 68 percent in 1999. There is a growing perception that much of aquaculture production leads to environmental damage. This is not correct for many aquaculture activities. Molluscs, seaweed and most forms of carp culture, for example, have only a very limited impact on the aquatic environment, and these species account for more than half of world aquaculture production.

However, it is generally recognised that the early rapid expansion of shrimp farming in countries around the Bay of Bengal, elsewhere in Asia, and in Latin America caused considerable environmental and social damage. This was in large part a result of unsuitable procedures for site selection, pond construction and pond management. Conversely, aquaculture itself has increasingly been negatively affected by coastal development, habitat loss, and non-point-source pollution, issues that are usually beyond the direct jurisdiction or control of fisheries management authorities.

Demand for Fishery Products

Demand for fishery products consists of two components: direct consumption of fish as food and fish used for animal feeding. Increasingly, fishery resources are also in demand as a source of leisure activities. Direct consumption of fish as food. Globally, fish provide about 16 percent of animal proteins consumed and are a valuable source of minerals and essential fatty acids. Regionally, these consumption figures vary, with fish accounting for an average of 30 percent of consumed animal proteins in Asia, approximately 20 percent in

Africa, and around 10 percent in Latin America and the Caribbean.

During the 1990s, global apparent consumption of fish increased. In 1990, apparent direct food consumption of fish, crustaceans and molluscs amounted to 71.2 million tonnes (live-weight equivalent). This figure increased to 97.2 million tonnes in 1999, an increase of more than a third over a period of only nine years, faster than the growth of world population. Thus average apparent consumption of fish increased from 13.4 kg per capita in 1990 to 16.3 kg per capita in 1999. This development was heavily dominated by events in China, which emerged as the world's largest fish producer during this period. Growth in Chinese fish production was so rapid that it overshadowed developments elsewhere. In fact, excluding China, the apparent consumption per person in the rest of the world actually declined from 14.4 kg in 1990 to 13.1 kg in 1999.

However, it is important to note that such global figures mask the very wide differences among countries in the level of fish used in food consumption. This ranges from less than 1 kg per person p.a. in some countries to over 100 kg per person in others. Moreover, in recent years the form in which fish has been consumed has changed. The volume of fishery products marketed in their fresh state has increased not only in absolute terms, but also as a share of all uses of fish. In 1999 slightly more than a third of all fish was marketed fresh against only one fifth in 1987.

During the 1990s, between 28 and 33 million tonnes of fish were used each year for the production of fishmeal and oil. Almost all of this was landed by marine capture fisheries. When the capture of pelagic fish cff the west coast of South America contracted as a result of the El Niño phenomenon, so did the production of fishmeal in the world: in 1998 only 23.9 million tonnes of fish were reduced to fishmeal and oil. By 1999 this figure increased again to 30 million tonnes or almost 24 percent of total world catch of fish, representing a return to a more normal level as a result of the recovery of fishing in South America.

Two types of users are increasingly using fisheries for leisure activities: active users such as recreational and sports fishers, and so-called non-consumptive and passive users such as tourists, sport divers and individuals who want to enjoy the environmental values of pristine marine environments (including fisheries). In many parts of the world, the number of stakeholders who want to use fisheries for leisure purposes exceeds the number of commercial fishers. Because all stakeholders, commercial or otherwise, have valid claims for access to living aquatic resources and have a stake in influencing the use of world fisheries, the emergence of this group is an important development. Because of their numbers and the fact that they are increasingly well represented and organised, these stakeholders in affluent developed economies have a growing influence on how fisheries can be used, including the use of fisheries for food production.

This new category of stakeholders has developed various strategies for effecting its objectives. In some instances, stakeholders have moved to prevent or otherwise modify exploitation of some valued species, either because the exploitation of these species had the potential to directly threaten species (e.g. marine mammals) or because of the effects capture fisheries could have on associated species, e.g. seabirds or seals. In other instances, they have asserted particular policy perspectives regarding the appropriateness of exploitation of certain environments or species and have worked to establish parks or protected areas in which certain fisheries activities are prohibited or limited.

In some cases such efforts have resulted in fishing methods being significantly modified, restricted and even prohibited as, for example, under the 1989 UN Resolution 44/225 on Large-Scale Pelagic Driftnet Fishing and its Impacts on the Living Resources of the World's Oceans and Seas and the related 1989 Wellington Convention for the Prohibition of Fishing with Large Driftnets (and Protocols).

Trade in Fishery Products

The gradual opening up of markets during the last decade has expanded international trade in fish and fish products in both volume and geographic coverage. Nearly 40 percent of all fishery production is now internationally traded. Around 80 percent of fish for human consumption ends up in three main markets. Exporting countries increasingly see trade in fish and fish products as a powerful vehicle for earning foreign exchange, especially through the export of high-valued species. For example, in Indonesia the government is aiming for approximately three-quarters of all foreign exchange earned from fish and crustacean exports to come from high-valued cultured species. Globally, gross earnings from fish exports by developing countries grew rapidly from US$5.2 billion in 1985 to US$15.6 billion in 1999, a level that raised fish export earnings to a level on a par with the total value of exports of such commodities as bananas, coffee, cocoa and rubber.

International trade is particularly important in fishmeal and oil. Some 65 percent of the volume of world fishmeal production enters international trade. This reflects the fact that fishmeal is produced primarily in

South America but is used as animal feed in feedlots in more wealthy economies elsewhere in the world. This growth in trade also reflects the global growth of the aquaculture sector, where fishmeal and oil are key feed ingredients. The increase in trade in fish and fish products in the 1990s has not been without difficulty. Stringent standards for fish imports have been introduced, mostly in the markets of developed countries.

Employment in Fisheries

Employment policies in the fisheries sector are seldom explicitly elaborated. Where they are explicit, they frequently concern the participation of local fishers as crew in a joint venture or on foreign flagged, licensed vessels fishing in the national exclusive economic zones (EEZs). Despite this, other developments in marine policy have had an impact on fisheries employment. For example, the introduction of the EEZs early in the 1980s affected fisheries employment. Employment in long-distance fishing fleets changed when countries lost access to fishing grounds. This affected both those that already existed, and those developed during the 1950s and 1960s by countries such as Japan, the former Soviet Union and several countries in Europe.

Some countries, particularly Spain and Portugal, relocated much of their activities to other nations' fishing grounds through fishing agreements. Other countries, particularly developing countries in Asia, expanded employment in fisheries, in part through effective occupation of their own fishing grounds in their newly created EEZs, but also by engaging in high sea fisheries. Still other countries worked directly to reduce the numbers of those employed in high seas and other fisheries. For example, from 1980 to 1995, the number of active Japanese marine fishers was reduced from some 457 to 278 thousand.

Although information on fisheries employment is only fragmentary, the available data show that during the 1990s employment in capture fisheries and aquaculture continued to contract in the developed economies and to expand in the developing countries. Already in the early 1990s, more than 90 percent of those fully employed in the fishery sector were employed by either developing economies or economies in transition. At the same time, fisheries employment in developed economies fell in capture fisheries, in part compensated by the slowly expanding employment in aquaculture. Overall, growth in employment in fisheries is slowing down significantly.

When fisheries are used as a means for providing social safety nets or for last-resort employment opportunities, such employment policies are rarely explicitly enunciated. For example, there has been an increasing tendency to allocate inshore areas with fish resources to small-scale fishers. This can be seen as an indirect way of preserving local employment opportunities and a source of food for local fishing communities. However, such policies rarely provide legal rules and mechanisms for controlling the inflow of fishers from other areas or communities. As a result, unrestricted catches by unregulated numbers of fishers mean that inshore fisheries are increasingly unable to render the benefits they were intended to provide, particularly when there are problems in other sectors that drive people into fisheries as an activity of last resort.

Developments in Fisheries Production

Developments in the global fisheries picture will to a great extent be determined by developments in fisheries management, which will have to face problems such as the increasingly contentious issue of access rights allocation. Overall, production from wild capture fisheries is approaching its biologically sustainable limit for many fish species. The bulk of further production increases will therefore have to come from aquaculture. However, aquaculture is at present also constrained to the extent that it relies for feeding purposes on fishmeal and fish oil that come from capture fisheries. This constraint may be overcome when cost-effective substitutes for fishmeal and fish oil emerge.

Beyond these constraints, future fish production increases will depend on further development of post-harvest utilisation technologies, and on successful resolution of various negative externalities such as habitat destruction, bycatch and pollution from intensive aquaculture. The future for the aquaculture sector varies by region. In Asia aquaculture is long practised and well established. It is expected that aquaculture will continue to grow and supply important segments of fish markets in Asia, increasing its share in total fish production.

The aquaculture industry will respond to the growing demand for fish by increasing production of already domesticated species, by domesticating additional species, and by expanding marine ranching. In doing so, the industry will need to resolve issues such as how to maintain genetic diversity of domesticated species, how to develop technologies for new species, how to resolve a looming shortage of feed for cultured fish and shrimps, and how to deal with pollution issues.

Aquaculture is an ancient activity that was usually carried out in harmony with other uses of the environment, but few of the old ways of raising aquatic animals and plants still prevail. Modernised technologies have resulted in substantial reductions in production costs for several species, but they have also created negative environmental externalities. Aquaculture will be increasingly subject to regulations ensuring judicious use of culture technologies and site selection.

The development of technologies, and a reduction in the costs of dealing with the undesirable environmental effects of aquaculture installations and intensive feed regimes, will allow these problems to be much reduced, if not overcome. Environmental considerations may also lead to the shift of aquaculture installations away from the coastal zone into offshore and submergible installations at sea, and in densely populated areas inland, into intensive recirculating systems.

There is concern in the aquaculture industry that the limited availability of feeds could constrain the expansion of aquaculture production. Aquaculturists using semi-intensive and intensive culture systems for fish and marine shrimps are particularly dependent on feeds that incorporate fish oils. Fishmeal producers expect that within a decade or so, the aquaculture industry will use up to 75 to 80 percent of all fish oil produced, and about half of the available white fishmeal. Although considerable research efforts have been undertaken, a satisfactory replacement for fish oil has still to be found. The problem of finding new fish feeds will soon become more pressing, given that semi-intensive and intensive culture systems for shrimps, marine fish and increasingly for carp and other freshwater fishes will expand.

Despite boom-and-bust cycles of overexpansion, market gluts and falling prices in northern Europe and North America, it is likely that the culture of salmon, sea bass, bream and turbot will still continue to expand, particularly in Europe. But the rate of expansion is likely to decrease and stabilise, probably at a few percent per year in terms of the volume of production. The rate of expansion will depend on the intensity of market promotion efforts, particularly in non-traditional markets.

However, there are cultured species such as carp, tilapias and catfishes that have much larger "natural" markets. For example, carp account for about 70 percent of world cultured production of finfish although they are produced in only a few countries. It is plausible to expect continued growth in traditional markets. Total production may well double by 2015, with growth slowing thereafter. If non-Asian markets could be found for Indian and Asian carp, production could grow more quickly.

After several decades of promotion, the culture of tilapia (mostly Oreochromis nilotica) took off at the end of the 1990s, not in its natural home of Africa, but in Asia, where it has found ready markets. It is also finding markets in Europe and North America. It seems most likely that tilapia culture will expand in Asia, intensify in Latin America and take off in Africa. The sector has been growing in volume by over 10 percent p.a. at the end of the 1990s, and it is likely to maintain that rate at least for the next two decades. Tilapia is a relatively inexpensive fish, and it can be used as an ingredient in processed products. By 2015, it is plausible that there will be a production volume in the order of 3 to 5 million tonnes p.a., increasing to possibly twice that amount by 2030.

Crustaceans account for about 5 percent of world production of fish and shellfish. While the culture of crustaceans has progressed steadily in volume, this has not always been the case for individual species. Given the growing demand for this luxury product and an increased ability to deal with disease (a severe problem in the culture of tropical marine shrimps), it would seem reasonable to expect that the culture of present domesticated crustaceans will expand as it did during the last few years of the 1990s.

Molluscs, foremost among them oysters, account for about one-third of world production of fish and shellfish. Recently, the production of mussels has stagnated. However, efforts are under way for these products also to find new markets and these are likely to be successful. Production of mussels can be expected to grow through new offshore technologies. Thus mollusc production is likely to resume its pace of growth.

Because of the high costs and long periods involved in achieving controlled reproduction and survival of the offspring of newly domesticated species, most recent domestications have been achieved in developed countries, in particular along the shores of the North Atlantic. This is likely to continue in the future. Production of species such as char, cod, halibut, coryphena and possibly large tunas and turbot is likely to grow and eventually stabilise. The level at which this will happen is largely determined by production costs vis-à-vis those in capture fisheries. For halibut, cod and tunas, which lately saw relatively high volumes in capture fisheries, cultured production will eventually be quite high. For Atlantic cod it may be several millions of tonnes. The technology is being developed for some types of halibut, production of which may increase quickly and reach hundreds of thousands of tonnes before the industry reaches serious market problems.

Capture fisheries of char and turbot amount to only a few thousand tonnes per year. So, in terms of volume and number of producers, growth of aquaculture may be slow for these species. Among the species mentioned above, Atlantic cod is perhaps the most interesting. Once this species has been domesticated and a commercially viable culture technology develops, the impact could be considerable not only on codfish fisheries and their markets but on the white fish industry as a whole. It took about 15 years for the Atlantic salmon industry to reach a yearly production of half a million tonnes and, as noted above, the marketing challenge was considerable. It is likely to take less time for the cod culture industry to reach that figure. It will start from a much better market position.

Total capture fishery production of Atlantic cod is about 1.3 million tonnes p.a. Atlantic cod has a market also in southern Europe and in Latin America, and the species can be sold in many product forms. Thus, assuming that commercially viable technology for the culture of cod is developed soon, by 2015 cultured cod production could be at least 1 to 2 million tonnes p.a. At present, of the major fishing nations, only Japan is engaged in sea (or ocean) ranching on a large scale. Over 80 species are being ranched or researched for eventual stocking in marine, brackish and fresh waters and, as a result, the population has increased for a number of species. Many other countries are also active in the enhancement and ranching of aquatic species, but most of them stock only a few marine species. Salmonids are the most widely stocked group of fish.

The future of sea ranching will depend on a number of crucial issues. One of these is how to deal with the externalities of ocean ranching and the potential ecosystem-related imbalances that it could create. Another is the ownership of the released animals. As long as ownership is not legally defined and technically impossible to defend, the private sector and governments will be reluctant to spend resources on maintaining ranching programmes. It is possible that new technology will permit the identification of released animals and their separation on capture. But, before sea ranching can contribute significantly to world fish supplies it must be integrated into an overall fishery management approach that includes habitat protection.

In inland fisheries, enhancement of fish stocks has been practised successfully in Asia, particularly in China. There is considerable potential for stocking and harvesting fish in both permanent and seasonal freshwater bodies in other parts of the world. The obstacle to rapid development of inland fish stocking is not technical, but rather financial and institutional. Most small freshwater bodies have a large number of users with long-standing traditional usage rights. Enhancement practices must be compatible with these rights and include mechanisms for perpetuating them. No doubt the practice of stocking fish in such bodies will spread, but the spread will be slow in Africa, and is not likely to receive priority in Latin America (with some notable exceptions). In Asia the practice will continue.

Wild capture fisheries production in the longer term will be influenced by advances in harvesting technologies and marketing developments. Estimates of total sustainable production still refer to the figure of approximately 100 million tonnes p.a. This is higher than the annual catches of 80 to 85 million tonnes of the 1990s because it assumes efficient utilisation of the stocks in healthy ecosystems where critical habitats have been conserved. Moreover, this estimated potential yield includes large quantities of living marine resources which thus far have been little exploited. Of these the best known are krill, mesopelagic fish and oceanic squids. The major factor affecting wild capture production is whether fisheries management can ensure that fish are harvested in a sustainable and economically efficient manner.

Clearly, technical developments will continue to increase the efficiency of fishing. Navigational technologies, electronic fish-finding devices, freezing technologies and gear improvements have already increased fishing productivity substantially. As international awareness increases about various fishing practices, more selective fishing gears that can reduce the catch of non-target species are being developed and spread. In addition, social pressures will continue to drive innovations to make use of unwanted but unavoidable catch. Nonetheless, technical gains will probably not fully compensate for management strategies where these are increasingly economically inefficient.

Market changes will also drive changes in technology. There is a growing trend to consume fresh fish instead of processed fish products. Increased access to markets, lower costs of the technology for delivering fresh fish, and price premiums paid for fresh fish are creating incentives to make this shift. The advent of labelling schemes further supports market changes. For example, labelling products by country of origin, and use of ecolabels recognising sustainably harvested fish enable producers to differentiate their products and take advantage of changing consumer preferences. However, these developments will not necessarily alter total production, just the form in which fish are produced for food.

The increasing competition of fish substitutes will force wild capture fisheries to deliver products that enjoy price premiums. For example, at the beginning of the 1990s, imports of farmed tilapia (fillets) and supply of pond-cultured catfish were believed to depress prices of white fish in the United States market. Similar situations have been observed in south European countries with respect to European sea bass and sea bream. In 1991 some 3800 tonnes of European sea bass came from culture, or just under 40 percent of the total supply. By 1999 aquaculturists produced more than 41000 tonnes of the species, accounting for 85 percent of the total supply, and sea bass prices dropped. Similarly, the sheer volumes of cultured European sea bass, Channel catfish and Atlantic salmon that are now available have affected prices for these and other substitutable food sources.

Fish food products. Growth in food demand for fish will not be uniform, and there will be significant differences between regions and countries. Delgado et al. suggest that this increase in meat consumption is expected to occur in parallel with a small decline in the real world market prices of both meat and grains. If these projections prove correct, fish consumption could still increase in developing countries where meat is not a common substitute for fish. Although consumption of poultry and pig meat is expected to continue to increase in the developed countries, this is unlikely to have a major effect on trends in fish consumption in these countries, where demand for fish is more influenced by variety and health considerations than for reasons of economy.

The world average per capita fish consumption grew from 9.4 kg in 1961/63 to 16.3 kg in 1999, or by over 70 percent. A study for FAO, taking into account per capita income growth and trend factors, estimates that per capita fish consumption could reach 22.5 kg by 2030, an increase of almost 40 percent over 1999 consumption. Total world demand would then reach 186 million tonnes in 2030, or almost 90 million tonnes more than in 1999. However, the plausible developments in capture fisheries and aquaculture discussed above may limit the growth in fish supply.

Consumption will continue to increase thereafter, but at a slightly slower rate, reaching between 30 and 35 kg per capita in 2030. However, in terms of total consumption, this increase is relatively small because populations in these regions are expected to grow only marginally. In Japan, the market is considered to be saturated, and it is likely that the total volume of fish consumption will decline slightly.

The 1999 average per capita consumption in the region was 17.4 kg. China's per capita consumption was much higher (25.6 kg). Demand is expected to continue to increase so that per capita demand in some countries in this region could reach about 40 kg by 2030. China's very rapid growth in per capita fish consumption will continue, as elsewhere in Asia, although this rate will begin to slow down over the next 30 years.

Since South Americans are large consumers of red meats, and are likely to remain so, per capita fish consumption is relatively low (9.6 kg in 1999), especially compared with North and Central America where per capita consumption is in the order of 16.5 kg. Furthermore, although per capita consumption in South America could grow slowly over the next 30 years, it is not likely to exceed 15 kg by 2030.

Demand for fish for animal feeding purposes. Small pelagic species and macro-algae have traditionally been used for animal feeding purposes and also as fertilisers. The use of fish and algae as fertiliser has almost ceased but vast quantities of small pelagic species and small volumes of fish offal are still regularly converted into fishmeal and oil, the bulk of which is used as feed ingredient. Fish used as raw material in the manufacture of fishmeal accounted in 2000 for about a quarter of total fish production (30 out of 130 million tonnes).

Thus far, except for small volumes of offal, all raw material for fishmeal and oil has been supplied by capture fisheries. In all likelihood this will continue to be the situation over the projection period. However, the competition for pelagic fish, a limited and highly fluctuating resource, will become more intense, and it is unlikely that the fishmeal industry will have the economic means to continue to purchase the 30 million tonnes of fish p.a. for reduction. The industry will need to find other raw materials, the most plausible of which could be zooplankton or krill. Finally, the onboard processing of unavoidable bycatch and offal into fishmeal may expand.

Demand for fisheries in leisure activities. Recreational and sports fishers are becoming more numerous than commercial fishers in North America, northern Europe and Oceania, and the number of leisure fishers will continue to grow around the world. These stakeholders will increasingly obtain guaranteed access to particular stocks and fishing grounds, sometimes regulated by managing fisheries agencies and sometimes by purchasing access rights from commercial fishers. Their impact will be felt in both freshwater and marine fisheries.

Indeed by 2030 the economic importance of recreational and sports fisheries could easily be greater than that of commercial fisheries for several species and waterbodies. This is a plausible scenario not only for developed countries, but also for "big game fishing" and similar activities occurring in tropical areas and on high seas. If access is reallocated in particular fisheries there will be an impact on revenues of the commercial fishing industry, even if the volumes captured in leisure fisheries are likely to remain comparatively small.

There is also a growing demand for "non-consumptive use" of living aquatic resources in developed countries. This demand for access to pristine and exotic resources has found indirect support in a number of United Nations conventions and agreements. The influence of non-consumptive user groups has also increased in less developed countries. It seems likely that demand for non-consumptive uses will continue to grow gradually. However, this is not expected to result in any significant reduction of commercial capture fisheries, particularly in developing countries, unless switching to such activities creates alternative livelihood opportunities that address local food security issues.

Fishery Trade Development

In coming decades, the world's trading system is likely to be further liberalised under WTO auspices. Trade barriers for fisheries products will eventually be lowered or even dismantled. In anticipation of these developments, fish processors are investing in low-cost countries to take advantage of cost differences. Regional specialisation in fish production means that some fish-processing industries will probably remain in industrialised countries, even if they have to rely on buying raw material from distant suppliers.

The expansion of international markets can be considered as useful but also problematic. For many communities, especially in developing countries, fish play a very important role in food security and constitute an important part of the diet. However, if fish producers or processors take advantage of access to distant markets for their fish in order to receive a higher net income, the local availability of fish may be reduced. Thus, increased trade can provide a mechanism for economic development, but local communities also need to change to accommodate it.

Carp production in China and India is very important but is not expected to expand greatly beyond existing markets in Asia, largely because the demand is limited by regional consumer preferences. In Africa and Latin America the situation is somewhat different in the sense that aquaculture constitutes a very small share of local fish supplies. In these regions abrupt increases of aquaculture fish production for local markets are not impossible, but at present seem unlikely. Thus, if a sudden increase in aquaculture production were to take place, it would most likely be directed at foreign markets. The direct effect on supplies in local markets would likely be small, with little competitive impact on local capture fisheries producing for local consumption.

Management of Fishery Development

The management issues that are receiving increasing attention in the fisheries sector fall under the headings of maintenance of biodiversity, ecologically sustainable development and, more recently, ecosystem management. These are important considerations for many stakeholders, all of whom are increasingly vocal about what constitutes appropriate practices and acceptable uses of aquatic living resources.

However, there is a largely unaddressed, yet fundamental issue underlying these concerns. It is that in spite of their intrinsic renewable nature, capture fisheries resources are essentially finite. And, as finite resources, they can only be exploited so much in a given period, beyond which overexploitation will cause them to produce less in the longer term, if not entirely collapse. This simple fact means that total wild capture fisheries production cannot increase infinitely, that these exhaustible resources have to be treated with care, and that what is produced must be shared among those engaged in harvesting. The latter can either be done on an ad hoc free-for-all basis which probably is ultimately unsustainable, through administrative processes, or through the use of rights-based management systems.

The single most important development for the future of wild capture production is how best to contain total catches at sustainable levels while allocating catches. To date, with only few exceptions, the matter of allocating resources among producers and other stakeholders has been addressed only indirectly, and not very successfully. To achieve sustainable and equitable use, most fisheries stakeholders and administrators will eventually have to recognise the power of using instruments that create and enhance incentives, rather than administrative decisions.

Fisheries management strategies based on explicit and well-defined access rights will have to become more common, or otherwise management entities will be overwhelmed by allocation conflicts. To facilitate decision-making, management responsibilities will need to be devolved to fishing interests and communities, with increased participation by stakeholder groups. As part of this, frameworks that control access to capture fisheries need to be strengthened, and created where they do not exist. This will not be a matter only for local or artisanal fisheries—the governance of high seas fisheries will also need to be more explicitly regulated.

In developing countries, the need to design management strategies that can cope with exit and entry issues in small-scale and artisanal fisheries will become more pressing. Although some countries (for example, the Philippines and Sri Lanka) are working to confront this problem, the vast majority of countries are still struggling with how to do so. Not being able to cope with issues of mass entry and exit could have severe implications for the food security of the people involved. Eventually, it will become politically and socially imperative that alternative social safety nets are developed so that fisheries are not the sector of last resort.

Sustainability issues are also important in aquaculture. The early experience with modern semi-intensive and intensive methods showed that access to and use of water, seed, feed, etc. was perhaps not as low cost as initially envisioned. During the 1990s aquaculturists had to accept that access to sites on land and in water and feeding practices became increasingly regulated to reduce the negative externalities of some aquaculture technologies (e.g. intensive pond or cage culture).

In many regions restrictions on species transfers, site selection and feed use have become quite severe, and this has increased production costs. Where profit margins are large, some of these regulatory changes have altered, but not eliminated, aquaculture production. However, in many developing countries the application of such regulatory concerns is difficult. There is little incentive to comply unless the consumers demand it and are prepared to pay a higher price for fish products.

15

Policy for Conservation Agriculture

Conservation agriculture (CA) and related practices has already captured many of the effects of policy, or more generally government action, on adoption. Governments use macro-economic policy, trade regulations, input subsidies, or education and extension to alter the decision-making environment in which farmers choose one practice over another.

Influence of Policy on the Conservation Agriculture

Agriculture has been subject to considerable state interest and intervention over the past half-century, perhaps more than any other economic sector. While it is possible to overestimate the influence of policies in farmer decision making, there is increasing recognition that the provision of public support in the form of guaranteed output prices, input subsidies, deficiency payments, cheap credit, or disaster relief has encouraged and facilitated massive investment by farmers in production capacity expansion. This is because of its continuing trend towards larger and fewer units of production, regional and enterprise specialisation, more intensive soil tillage, increased reliance on agrochemicals, and in many locations, surplus output.

Many governments have introduced a variety of programmes to encourage the adoption of CA-type practices. With extension services, subsidies and taxes, these initiatives have achieved some important results. For example, the success in promoting CA practices in certain developing regions, particularly Latin America, is noteworthy, and policy has played an important role. Many of these stem not from government policy but from extraneous factors and local traditions. Indeed, many programmes promoting CA throughout the world have been relatively ineffective because of contradictory signals and incentives from existing subsidy programmes. For example, policies designed to promote sustainable agriculture can be undermined by other, typically richer, policy measures in support of highly erosive row crops such as groundnuts and tobacco, or by weak or slow-to-respond research and extension efforts.

Given the perceived environmental impacts over the past half-century, some have argued that the decoupling of agricultural support from production decisions would represent the most effective means by which governments could alleviate environmental degradation. There is debate concerning the means, both direct and indirect, by which governments can promote conservation in agriculture effectively.

In promoting CA, a key concern for policy-makers is whether CA provides a positive or negative net return to potential adopters. Once this uncertainty is rectified, Uri recommends:

- education and technical assistance where conservation is profitable but the farmer is not aware of the technology or its profitability, or does not have the skills to implement it;

- financial assistance where conservation is not profitable to the individual farmer but would provide substantial public benefits;
- long-term research and development;
- land retirement; and
- regulation and taxes where conservation behaviour is required of all farmers, or for those participating in related income support programmes.

With respect to the first approach, McNairn and Mitchell argue that encouraging the adoption of conservation practices requires assurance of long-term benefits from adoption; unambiguous, easily understood and accurate information; and the promotion of multiple economic and non-economic benefits. Education plays a key role in motivating adoption and requires tailored, credible, and appropriate information and experience that is communicated through the proper channels. Extension services to provide information and assistance can be highly effective, especially in the case of new or emerging technologies, although public agents need not be the exclusive providers of such services.

Financial assistance for the adoption of various conservation practices is well established in Europe and, to a lesser degree, North America. Assistance can take a variety of forms, such as tax credits on equipment, machine rentals, cost-sharing programmes and direct subsidies. Assistance is most suitable to help overcome significant initial investments and transition costs, and in cases where adoption is unprofitable from the individual farm perspective. That is, to overcome non-adoption because of onerous information demands, state support is useful.

A less interventionist policy approach might focus on research and development to enhance the benefits of CA adoption by improving performance or reducing costs. This approach relies on voluntary adoption and aims to increase the odds of this occurring by making the practice more attractive. However, research and development is a long-term policy strategy with an uncertain probability of success.

Land retirement is only suitable in instances where soil erosion concerns are so significant as to warrant conversion to permanent cover crops. Typically, this approach requires significant public financing to compensate farmers, and it is infeasible in areas highly dependent on a limited land base for the production of foodstuffs.

Finally, although tried in some locations, regulating soil erosion limits is not a common approach. This situation probably arises from political awkwardness and onerous enforcement/compliance demands. This is especially so where meeting a soil loss regulation through use of no-till results in significant declines in net returns. A more common regulatory approach involves cross-compliance measures whereby eligibility for a support programme depends on the adoption of certain conservation practices. Because compliance is by choice, programme implementation is liable to be more politically feasible and economically efficient. With respect to the use of taxes on soil erosion, it is possible to induce CA adoption and even pasture conversion. However, meaningful levels of soil conservation involve significant revenue losses. Hence, although possible, taxation is politically infeasible.

The inconclusive nature of empirical studies, and obvious site-specific nature of many results, suggests that a universal approach is not possible. In order to accommodate differences between farms, farmers and economic circumstances, a targeted policy approach may be preferable. In other words, policy mechanisms such as grants or extension services could be geared to the particulars of a location or, preferably, to individual farmers and their farm operations. While a targeted policy approach places a heavy administrative burden on policy-makers, it could achieve greater efficiencies than a more uniform approach, and may represent the most effective means of encouraging CA adoption.

Although a targeted policy approach may be most appropriate for the design of programmes directly promoting CA, there are some alternative policy prescriptions that may be more universally applicable. For example, Isham points out that parallel investments in social capital may be necessary to create a sufficiently enabling environment for the adoption of desirable project activities, and this may apply strongly in the case of CA.

Economic and Policy Analysis

Specialised policy and economic analyses are prerequisites for the appropriate design and correct targeting of CA policies. Policy analysts and economists interested in CA can make use of numerous new techniques and ways of thinking. Sustainability indicators are one example. These capture changes in farming practices that alter the sustainability of the farming system in some quantifiable way that conventional analysis may fail to capture. Therefore, sustainability indicators help describe the evolution of soil productivity over time or present its status in terms that better contrast conditions under CA and conventional management. Sustainability indicators are applicable at the local farming-systems level, at intermediate levels such as the community or region, or at higher levels. To the extent that more comprehensive sustainability measures incorporate these indicators, changes in farming practices will cause changes in the accompanying measures.

At the village and farm level, sustainability indicators assess the sustainability of specific farming systems and, by inference, the sustainability of soil tillage within a given farming system. There are several variables at the farm level that could serve as such indicators. Indicators that are more comprehensive define sustainability in an operational sense, using concepts such as sustainable income. This is the potential income that can be derived from resource use in perpetuity. In some cases, the indicators that accompany these definitions link farm-level soil degradation with national accounting techniques.

At the macroeconomic level, the system of national accounts has integrated soil degradation through formal green accounting initiatives such as the United Nations System of Integrated Environmental and Economic Accounting. Other national indicator approaches include the World Bank's calculations of genuine savings rates. These adjust net domestic savings for changes in the value of resource stocks and pollution damages while the Pearce-Atkinson indicator incorporates elements of the genuine savings idea. Indicators such as this can convey the message powerfully to decision-makers that soil degradation is resulting in a loss in national wealth, and so encourage greater efforts to promote more sustainable practices such as CA.

Analysts who have to assess the attractiveness of projects involving CA or competing farming practices can adopt a number of measures. Such efforts are important because some of the benefits of adopting CA do not show up in conventional cost-benefit type analyses, or in comparisons of CA and alternative practices in narrowly-defined financial terms.

Non-market Valuation Techniques

It is common practice to use non-market valuation techniques to incorporate the benefits and costs of farming practices that are not priced in markets. Examples include downstream siltation from soil erosion, or loss of organic fertiliser where dung is used as a fuel instead of on farm fields. The valuation practices most appropriate to comparisons of CA and conventional farming practices include replacement cost, changes in productivity, direct and indirect substitute approaches, preventive or mitigative expenditures, and hypothetical or constructed market techniques.

Soil Depletion

Economic analyses at the project level can incorporate the depletion of soil as a form of natural capital under conventional tillage practices, so enabling fairer comparisons with CA. This depletion constitutes a cost of non-sustainable cropping in addition to normal production costs. It is a user cost as it yields short-term gains at the expense of future income. Omitting user costs results in an overstatement of the net economic benefits of current cropping practices that deplete soils. Several techniques are available to calculate the user cost of depleting natural resource stocks. Two common approaches are the net price method and the marginal user cost method.

Whole-farm Budgeting

Proper environmental analysis requires the assessment of changes in environmental conditions in terms of the full range of behavioural responses that occur. When farmers adopt CA, numerous ancillary changes can be expected, such as crop switching, changes in pest control measures, shifts in cropping duties for household members (by gender), etc. For this reason, comparative analyses of CA and alternative practices should adopt a whole farm approach to capture the full range of these behavioural changes.

Alternative Project Evaluation Techniques

While project work makes universal use of cost-benefit analysis, other project evaluation techniques hold promise for the appraisal of CA projects or technologies. These include multi-criteria analysis (MCA), cost-effectiveness analysis, decision analysis, environmental impact assessment and participatory methods. MCA recognises that government decision-makers and smallholders have many objectives in mind when deciding about agricultural project viability and on-farm management practices, respectively; more than a cost-benefit analysis alone can capture. In addition, various trade-off techniques, such as trade-off curves or more sophisticated analytical techniques, can help assess the trade-offs amongst competing objectives.

Bibliography

Bartlett, M.S., "Properties of sufficiency and statistical test", *Proc. Roy. Soc*, 1937.

_______________., "Some examples of statistical methods of research in agriculture and applied biology", *J. Roy. Statist. Soc. Suppl.*, 1937.

Blaney, H.F. and W.D. Criddle, "Determining water requirements in irrigated areas from climatological and irrigation data", *USDA(SCS)TP* - 96, 1950.

Bwalya, M., "Conservation farming with animal traction in smallholder farming systems: Palabana experiences", In: Kaumbutho, P.G. Simalenga, T.E. (Eds.), *Conservation Tillage with Animal Traction*, ATNESA, Harare, Zimbabwe, 1999.

Calegari, A., Darolt, M.R. and Ferro, M., "Towards sustainable agriculture with a no-tillage system", *Advances in GeoEcology*, 1998.

Derpsch, R., "Historical Review of No-Tillage Cultivation of Crops", *JIRCAS Working Report*, 1999.

Ellis-Jones, J. and Mudhara, M., "Conservation tillage for resource poor farmers: the critical importance of farm power", In: Agro-ecological and economical aspects of soil tillage, *Proceedings of the 14th ISTRO Conference*, 27 July - 1 August, 1997, Pulaway, Poland.

Evans, J., Banda, A. and Seymour T., *Opportunities for better soil management*, Malawi, 1999.

FAO, "Soil conservation and management for small farms, Strategies and methods of introduction, technologies and equpment", *FAO Soils Bulletin*, FAO, Rome, 2000.

Fowler, R., "A hundred years of learning. Introductory remarks, Conservation tillage for sustainable soil management workshop", *Global Dialogue, EXPO 2000*, Hannover, 2000.

Gommes, R.A., "Pocket computers in agrometeorology", *FAO Plant Production and Protection Paper* No. 45, FAO, Rome, Italy, 1983.

Shapiro, S.S. and M.B. Wilk, "An analysis of variance test for normality", *J. Am. Statist. Assoc.*, 1965.

Veldhuizen, L.van, Waters-Bayer, A. and Zeeuw, H. de, "Developing Technology with farmers: A trainer's Guide for Participatory Learning, *Working Document*, 1997.

Index